5-1-81

I.A.L.

With thanks for the opportunities you gave me to learn the business and the trust you put in me as evidenced by my assignments with Guardsmith.

Best wishes to you and your family.

Gene Jannurn

SECURITY
SUPERVISION:
A Handbook for
Supervisors and
Managers

Eugene D. Finneran, CPP

BUTTERWORTH PUBLISHERS INC.
Boston   London

Library of Congress Cataloging in Publication Data

Finneran, Eugene D   1927-
    Security supervision.

    Includes index.
    1. Industry—Security measures.   I.   Title.
HV8290.F56   658.4'7   80-22801
ISBN 0-409-95025-4

Published by Butterworth (Publishers) Inc.
10 Tower Office Park
Woburn, MA 01801

Printed in the United States of America

This book is dedicated in loving memory of my father, Dennis C. Finneran, Sr., and with great affection for my mother, Mrs. Martha C. May.

CONTENTS

FOREWORD

Within the past ten to fifteen years America, as well as the rest of the world, has undergone major changes. Perhaps the area of security has changed as rapidly as the area of global environment. Security has certainly come of age; in the past it was a necessary evil usually delegated to other than the most qualified. All this is different now.

Mr. Eugene D. Finneran has literally kept abreast of the emerging profession, both as an active agent and as an educator of those who were assigned to work with him. Those of us who did not share with Mr. Finneran now have the opportunity, for in his work he moves the reader from the very basic to the potential of security managers.

In his work, Mr. Finneran recognizes the needs of both the individual without prior experience and the individual who has been placed in the role of security manager. He does this in a very easily understood and succinct method. Much to his credit he does not fail to advise the reader to proceed; indeed, he refers the reader to further avail himself of the resources each step along the way.

Here, then, is the text which may lead you to better understand and to work within the newly emerging professional world of security. Mr. Finneran's work will certainly open the path to the individual. He has given the student the key to the future world of risk management.

James T. Santor
Administration of Justice
Department
Clark County Community College

PREFACE

Our objectives in writing this book have been to provide an insight into the science of security as we have known it in the past, where it is today, and what we can expect in the future. We will be looking at security through the eyes of the security supervisor, the man in the middle. We want this book to be as valuable to the security manager as it will be to the guard aspiring to advance himself or the student seeking a degree in this rapidly expanding professional career field.

Just as learning the science of mathematics begins with simple addition and subtraction, rather than trigonometry or calculus, so the science of security needs to have an easily understood basic textbook. We have carried it a step further by providing information and techniques which will allow the reader to move into the arena of the loss prevention professional.

This book then will provide a text which will be useful to both the beginner and the expert; it will provide the security manager the means to evaluate the effectiveness of his present program and can be used as an effective guide for the training of a security guard force.

The security profession is ever increasing in importance to the well being of our nation. Today, security is very important; tomorrow, risk management will be essential if our economic system is to continue to function effectively. Business losses, directly attributable to criminal activity, have been allowed to flourish due to lax or non-existent loss prevention measures. Such losses are a major reason for the continuing rise in inflation since they cause a comparable rise to the consumer.

Adequate loss prevention measures can result in the reduction of product losses, permitting a reduction in product cost. For these reasons, security should be the concern of every man, woman and child in our world society.

For the most part this book deals with loss prevention programs as they would be encountered in an industrial or a wholesale commercial setting. However, the techniques, procedures and controls which are set forth can, in most cases, be applied to any setting where security is required. True, a subject such as vehicle control would not normally be the concern of the security supervisor responsible for security at a retail outlet, but shipping and receiving loss prevention procedures would be very applicable to this supervisor.

When we speak of security supervision, or the security supervisor, we include all supervisory positions in our profession. Although the supervisor of security guard forces accounts for the vast majority of the supervisory positions in our industry, we also employ supervisors, usually with specialized knowledge, in many other areas. Some examples would be supervisors of loss prevention programs for retail stores, hospitals, educational institutions, governmental programs, electronic data processing and financial institutions. Supervision is also needed in the areas of crowd control, shopping malls, airport preboard screening operations, residential community security, hotel/motel security, casino security, central station alarm companies, highrise building security and the security of construction sites. We must also provide for the supervision of our audit/inspection, consulting, and investigation functions. In many cases one supervisor is assigned responsibility for all the security-related functions within a given business.

Security, by definition, implies protection. The Security Industry is primarily in the business of protecting business assets. This protection, by necessity, is extended to the protection of the executives who manage businesses and the workers without whom there would be no assets to protect. Rising crime rates and improved technology are rapidly expanding the role of security into the protection of private residences. This protection, which in the past was only available to the wealthy, is being applied to homes of the general public in the form of smoke detection devices, central station and local burglar alarm systems, improved locking devices, and homes which are being constructed with greater security built into the design.

These changes and technological advances are requiring the professional security managers and supervisors to take advantage of continuing educational programs to keep abreast of and utilize changing techniques in our industry or fall by the wayside. For the student, about to choose a professional career, the expanding role of security provides a vehicle which can transport him into the board room of many diverse businesses. As we point out in Chapter 6, the professional security director or risk manager must have a solid working knowledge of all aspects of the business enterprise which employs him. In the future it will not be unusual for corporations to require their risk managers to have attained an MBA or CPA in addition to a CPP.

In Part I of our book we look at the question of selecting a proprietary or contract source of security service; the duties and responsibilities of the supervisor; the important elements of selecting and motivating personnel of the security force; what is involved in the organization of a security force; and identification and examples of security hazards and definition of risk management and loss prevention while outlining some techniques for implementing loss prevention measures.

Part II is concerned with physical security procedures and equipment; the importance of establishing a method of identifying and controlling personnel and vehicle movement; control of keys and use of locking devices; protective lighting systems; alarm systems; and access or entry control techniques, systems and equipment and the effective use of closed circuit television.

Part III looks at the prevention of fires and the equipment that is used in fighting fires; the importance of safety to the employee and the employer; and what is involved in planning for disasters and other emergency situations including the importance of knowing proper first aid techniques; with the handling of bomb threats and civil disturbances; and terrorism and its effect on executive protection.

Part IV deals with the conduct of security patrols; report writing; public relations as it pertains to the security function; and the determinations necessary for arming security personnel and the proper use of the hand gun.

In Part V we discuss the legal authority and problems facing the security department and its personnel and conclude by providing some guidance on the subjects and techniques that make up a training program for security personnel.

There are many more topics which could have been treated here, and in most cases, more detail could have been provided. However, to have done so would have defeated our purpose of providing an easily understood, basic insight into the complex field of security/loss prevention. We feel that what we have provided and the manner in which we present it will give the security practitioner a tool which can be put to good use in advancing the knowledge and proficiency of security personnel at all levels.

At the end of each chapter of this book we have included a list of questions for discussion and have pointed out the major considerations within the content of the chapter. Since it is our intention that this book be used both by the working security supervisor and manager as well as the student getting his first insight into this profession, we felt the inclusion of these pages was necessary. These questions will assist in the review of material presented in the classroom. They also can be used as review tests and administered during the training program to assess the progress of the security trainees. As we will point out throughout this book, the foundation of a good loss prevention program is a viable security system. The single most important element within that system is the security force. The effectiveness of the security force is directly dependent upon the knowledge and ability of the security supervisor. The success or failure of the system rests on these men and women.

Our thanks must go to the many people who gave us encouragement or advice which has made this a more thorough and a much better book. We would like to give special thanks to the following individuals whose contributions to our knowledge and experience made the preparation of this work possible.

To James T. Santor, Educator, who reviewed our efforts and has written the forward, we extend our deepest appreciation.

To Mark Lipman, President, Mark Lipman Services, who encouraged us to work on this book, we give our thanks and a warm regard.

To Walter M. Strobl, CPP, President, Strobl Security Services, Inc., Security Professional and Author, whose association greatly advanced our knowledge of the profession, we extend our thanks and best wishes.

To Ira A. Lipman, CPP, President and Chairman of the Board, Guardsmark, Inc., who provided us the opportunity to learn and display our abilities in the areas of business management, security consulting, managing technical services and security training programs, we would like to express our appreciation.

To Louis Charbonneau, former Managing Editor, Book Publications, Security World Books, and Greg Franklin, Acquisitions Editor, Butterworth Publishers, Inc., who provided us with thorough critiques and recommendations which have greatly improved the end results of our efforts, we add our sincere thanks.

Our greatest thanks must go to Ms. Nila M. Bieker, a security professional in her own right and my loving wife. Without her help and encouragement, this book would never have been completed.

One of the greatest changes that has taken place in the field of security has been the introduction, into this traditionally male-dominated profession, of female officers, supervisors, risk managers and directors of risk management. We want it known that we recognize the important contributions made to the professionalism of our field by the female professionals among us. In our early manuscripts there were many he/her and his/hers, so many in fact that it became cumbersome and difficult to deal with. In the final manuscript we have decided to use the traditional masculine gender with the understanding that wherever such references appear, we are referring to all professionals in our industry, male or female.

I. GUARD FORCES: THEIR SUPERVISION AND PURPOSE

Chapter 1

THE DECISION—PROPRIETARY OR CONTRACT?

In the early years of World War II, even before the United States was an active participant, our government began to rebuild our military arms and equipment productive capabilities. The War Department negotiated contracts with civilian manufacturing companies to develop and produce military equipment from weapons and ammunition to bombers and troop transports.

It was recognized that plans, blueprints, specifications and much of the finished product would be classified as military secrets and would require special protection. The War Department stipulated that companies engaged in war production would be required to meet certain minimum security requirements. Among these was the establishment of security guard forces. The emphasis was on the prevention of sabotage and denying foreign espionage agents access to our military secrets. The guard forces which were established were primarily proprietary. That is, the members of the force were direct employees of the protected firm. As will be repeated throughout this book, the guard force, although only one element of the security system, is the most important element of the system.

Many of the firms, whose contact with security came originally with the granting of government contracts, have stayed with the proprietary concept of employing a guard force. Other businesses have chosen to obtain these services by contracting with a company that specialized in providing guard personnel and security services. Still other businesses utilize a combination of the two options. Careful consideration should be given to the needs of a particular organization before a decision is reached to utilize either proprietary, contract or a combination of the two.

Although neither the security officer nor the supervisor will normally be concerned with selecting the method of obtaining a guard force, they should be familiar with the benefits and drawbacks of using either system.

After all, most of them would not have selected this career field if they did not intend to progress into management at some future point.

PROPRIETARY GUARD FORCES

Most of the benefits to be derived from the use of a proprietary security force form the basis for the biggest drawback to the use of this system. That drawback is cost. In order to be effective the proprietary force must be a part of an established professional security department. In Figure 1-1 the relationship between the corporate director of security (risk management) and other directors or officers of the corporation is shown. It should be noted here that as the role of security in our society changes so do the titles and terminology used to describe the various positions. In many businesses today we no longer have a position of director of security, instead we have director of risk management or director of loss prevention. The same changes are seen in security structures of individual facilities or facilities which come under the corporate umbrella. At this level the titles are usually risk managers and loss prevention managers. These titles more appropriately describe the duties of today's security departments, directors and managers.

Figure 1-2 depicts the structure of the security department at the corporate level. It should be pointed out that no two corporations necessarily use the same organization. In some corporations the director of the security program has no staff other than administrative and his function is primarily advisory in nature. Other corporations will have a more extensive staff than depicted here. The illustrations are provided solely for the purpose of showing the relationships and normal staff requirements for an organization of this type.

Figure 1-3 depicts the relationship of the guard force to the risk manager and the other sections within the security department. The staffing of this particular department has been overstated to show the necessary functions within the department. In reality, many of the duties cited would be assigned to one or two assistant managers or supervisors.

The advantages of using a proprietary guard force and security department, which were briefly mentioned earlier, are: the ability to pay higher wages and provide greater fringe benefits; the members of the security department owe their loyalties to the using organization, their employer; better control of employee assignments; better control of employee selection and background investigations; and a lower rate of employee turnover.

The disadvantages of using proprietary forces, as previously stated, begin with the cost—not only the higher cost of personnel procurement, wages and fringe benefits, but also the cost of uniforms, providing vacation and emergency relief (usually at overtime), maintaining the supervisory and

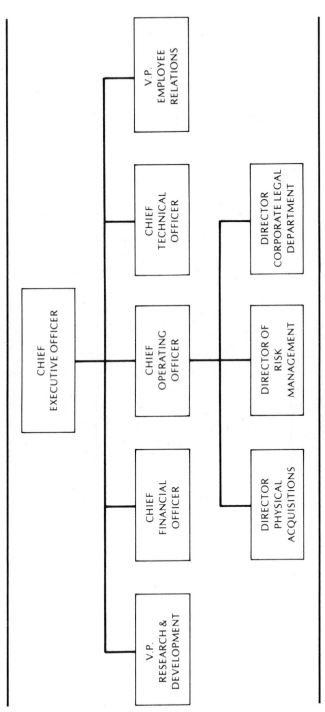

Figure 1-1. Relationship between the corporate director of security (risk management) and other directors or officers of the corporation.

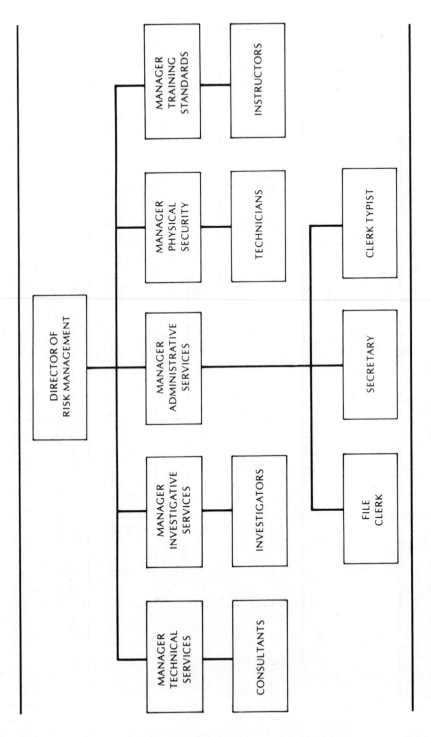

Figure 1-2.   Structure of the security department at the corporate level.

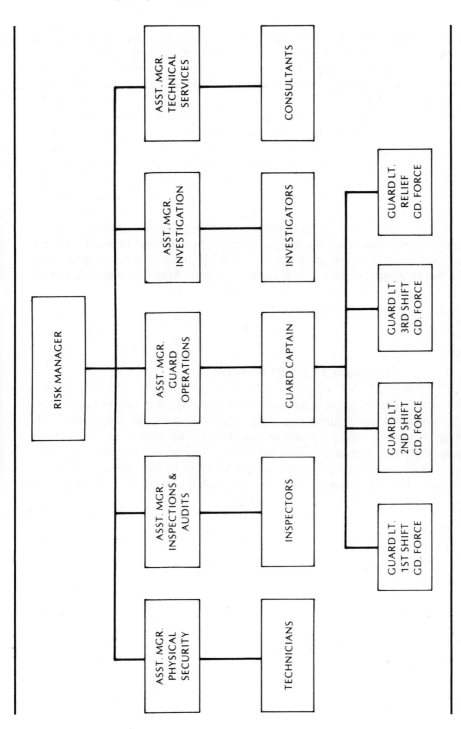

Figure 1-3. Relationship of the guard force to the risk manager and the other sections within the security department.

management staff, and payroll taxes, unemployment compensation, workers compensation, liability insurance and social security contributions.

If a proprietary guard force is to be employed, employment standards which specify educational level, experience, reputation and physical fitness requirements will need to be set up. Background investigations should be mandatory and should include a check of police files, where permitted by law, for any record of arrest/convictions. Polygraph or psychological stress evaluation examinations are helpful, but some jurisdictions do not allow them to be used as a bar to employment. Provisions must be made for uniforms and for obtaining relief personnel once the proprietary security force becomes eligible for vacations, as well as for medical and other emergency relief.

Companies which use a combination of proprietary and contract guard forces and security services usually employ the basic guard forces and contract for vacation relief, emergency coverage and additional coverage when needed for special events, construction projects and so on. Many companies also contract for investigators and consultants rather than maintaining such specialists on the staff of their security departments.

CONTRACT GUARD FORCES

Figure 1-4 provides a good indication of the savings that can be realized by contracting for guard forces and other security services. By comparison with Figure 1-3, all of the functions of the security department can be accomplished by the contract company with the exception of the using organization's manager or individual who has been delegated to oversee the security function. In many organizations this responsibility is given as an additional duty to the personnel manager, maintenance manager or other responsible person within the organization. The responsibility for inspections and audits may remain proprietary or can be contracted at the option of the user. Note that the user has an added benefit in that he has a reserve force available through the contracting company which may be used in case of disaster, civil disturbance or at any time the need might arise.

The advantages of utilizing a contract guard force begin with the savings which can be realized. The contract company recruits, screens, pays, trains, uniforms, insures, schedules, replaces and supervises the guard force. Consultants, investigators, inspectors and auditors can be contracted for on an as-needed basis rather than maintaining such personnel on the payroll. In many cases the contract company will provide expertise and advice free of charge to their guard force clients. The average contract guard force supervisor or manager has a broader experience base on which to draw, due

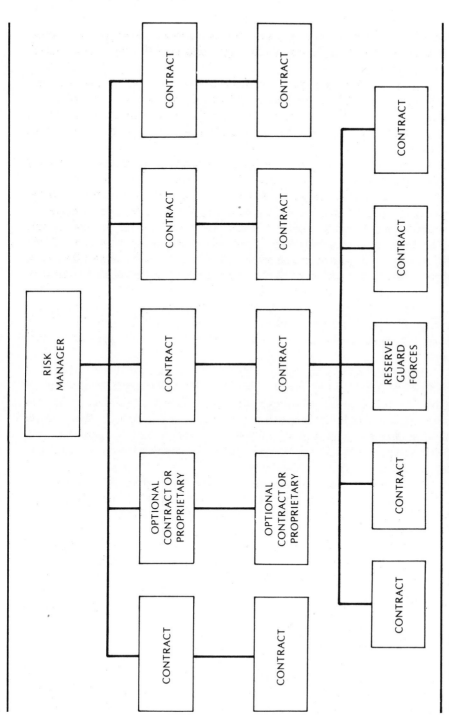

Figure 1-4. Structure of the security department by contracting for guard forces and other security services.

to the variety and number of guard forces he must supervise. A problem which has been solved for one client might also benefit other clients served by that organization.

Figure 1-5 depicts a possible corporate structure of a large contract security company. These companies will usually provide advisory guidance, budgetary limits and goals, and administrative assistance at the corporate level, and set policies, while leaving the day-to-day operations to the group offices.

Group offices, in turn, delegate operating procedures to regional offices which oversee the numerous profit centers under their control. To make this more readily understandable, picture a group office in each of the major geographical areas of the country, i.e., the East, Midwest, Southeast, Southwest and West. Under each individual group would be four or five regional offices strategically located for control of the profit centers. Profit centers would be located in the major cities of the region or in other areas where a sufficient number of clients require offices to be located to provide the necessary support to client operations.

Client support services, such as consulting, audit inspection and investigative services, will normally be available at the regional office level. However, some profit centers will also be capable of providing these services.

There are, of course, many contract services available throughout the country which are locally operated and which provide some of the services offered by the large corporations. These firms, due to their size and limited overhead, are able to offer guard force services at a lesser cost than can the major contract organizations. By the same token, they do not have the resources to provide all of the services and expertise which are available through the major firms in the industry.

If a determination has been made that an organization will use contract services, rather than proprietary, there are some guidelines which should be followed to insure receiving the best service available for the security dollar spent. If evaluation of needs determined that the only service required is the manpower necessary to staff the guard force with limited outside supervision, and special coverage for strikes or other civil disturbances is not anticipated, the best source of service will probably be one of the locally owned security guard companies. On the other hand, if it is determined by an organization, or by a consultant, that a total security/loss prevention program is required, then services should be acquired from one of the larger, nationally operated companies.

In either event, a list of businesses which are presently being served by the profit center or local business which would service the facility should be requested as part of the proposal to provide the service. The list should include clients which have used the contract service for a short period of

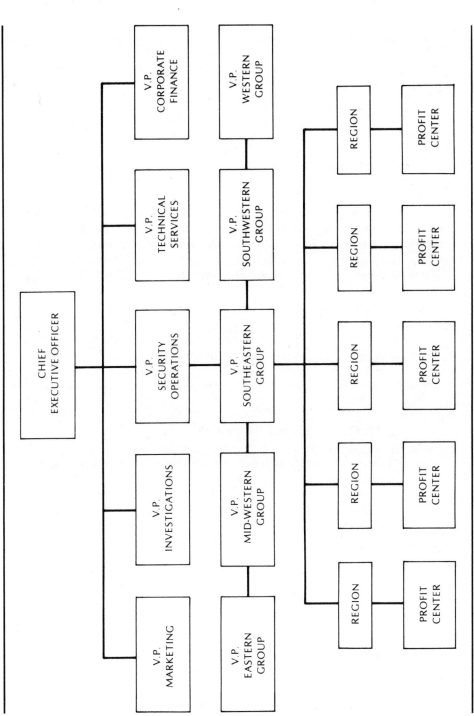

Figure 1-5. Corporate structure of a large contract security company.

time as well as long-term customers. By checking with the former, problems, if any, that might arise in the "start up" of service to the company can be determined. From the latter, the type of continuing service that might be expected can be determined.

Experience and background information on the management and supervisory personnel who will service the account, as well as a personal interview with these individuals prior to contracting for the service, should also be requested. They should be allowed to tour the facility and point out any known trouble areas or areas requiring special consideration and should then submit their opinions on staffing of the guard force and handling of special situations.

Companies which employ proprietary guard forces should consider contracting with a security firm for vacation relief and emergency or short-term coverage such as caused by sickness or family problems or to staff a temporary post such as a construction gate. For the short-term assignments where little advance notice is given, the contract rates will be considerably higher to cover the overtime that such jobs require. However, in most instances the rates will still be less than paying the proprietary guards at their overtime rate.

In Chapter 4, we will discuss the organization of a guard force, after a decision has been made as to the selection of the type of force to be used.

QUESTIONS FOR DISCUSSION

1. What is the difference between proprietary and contract guard forces?
2. Name some of the benefits of using a proprietary guard force.
3. Discuss the major drawback in establishing a proprietary guard force.
4. What are the benefits of selecting a contract service to provide the guard force?
5. Why do the terms loss prevention management or risk management more appropriately describe the duties of today's security departments?
6. What factors should be considered in selecting the type of guard force to be employed?
7. Should a company that employs a proprietary guard force consider using contract guards to supplement that force?
8. Discuss some of the services other than providing a guard force which are available from contract companies.

9. What is the advantage of contracting with a locally operated contract guard service rather than the larger national service company?
10. What services and advantages can be obtained from a nationally operated company that probably would not be available from the smaller local service company?

MAJOR CONSIDERATIONS

1. The role of security is no longer limited to organizing and supervising a guard force. Today, the director of the security function must be concerned with all aspects of loss prevention.
2. The decision to use a proprietary guard force, a contract guard force, or a combination of both must be based solely on the needs of the using organization as determined after careful study.
3. Management's recognition of the importance of risk management will determine the security budget. The budget will be a major determining factor in deciding the method of employing the guard force.

Chapter 2

THE SUPERVISOR

Ideally, the ratio of supervisors to employees may be as low as one super-visor for every three employees. In the security industry, the supervisor responsible for investigation or consulting may come close to that ratio but the guard force supervisor will rarely enjoy a rank structure which will permit such close supervision.

LEVELS OF SUPERVISION

It is important to determine where supervision begins within the security guard force and how far up the ladder it goes before the position is identified as a manager. Guard forces, like police departments, are quasi-military organiza-tions. Supervisory positions are found in both, identified by rank titles, of at least Sergeants, Lieutenants and Captains. The captain will normally be the last uniformed supervisor unless the force is exceptionally large which might justify the position of Major or in rare occasion, Colonels. In Figure 1-3, Chapter 1, an assistant manager for guard operations, a captain and four lieutenants were listed.

In that structure the assistant manager, although responsible for all guard force operations, has only one person he personally supervises—the captain. The captain has four lieutenants which he directly supervises. The lieutenants, depending upon the size of the security force, may have one or more sergeants reporting to them. In this case the sergeants would exercise supervisory authority over the other members of the guard force assigned to their shift.

Few security forces will enjoy the luxury of this extensive a command or supervisory structure. In most cases the guard force supervisor, reporting to the manager, will be a captain or lieutenant without the intermediate position of assistant manager. The supervisor may or may not have shift supervisors to assist in the supervision of the security guard force. Rarely will shift supervisors have another line of supervision between them and the personnel assigned to their shifts.

Many security guard forces will include sergeants, who have been given the rank as a reward for long and faithful or exceptional service and who exercise no supervisory authority over other security officers. This practice may be justified as a motivational tool, but it does not enhance the prestige or authority of those personnel required to supervise the activities of the members of the guard force.

RESPONSIBILITIES OF A SUPERVISOR

Now that we have some idea of what levels of supervision might be found within a security organization, we will examine just what it is that a supervisor is required to do. A supervisor, not only in a security department but any supervisor, is assigned certain tasks by management and is made responsible for the completion of those tasks. He is assigned the necessary tools and manpower to carry out the task and he must be given sufficient authority to permit him to complete his tasks successfully.

It sounds very simple and very easy. If, for example, a supervisor is assigned ten men with five picks and five shovels and is told to have them dig a 10' by 10' hole in the exact center of the west parking lot, he could find the task simple, unless of course the men could not agree on where the exact center of the west parking lot was located or who was to use the picks and who was to use the shovels. In this situation the supervisor would have to start earning his supervisory pay. He would be called upon to make decisions such as where to dig and which of the men were to swing the picks and which were to wield the shovels. If the men did not agree with his decision, he would be called upon to cause them to accept his decision, like it or not, by exercising his authority and leadership ability.

Of course, he could always fall back on the old crutch of poor supervisors, mutter to himself, "If you want something done right you have to do it yourself," give the men a break, and start digging. He might even try to justify this type of action by saying he is setting the example, showing the men how it is done or that he would never ask his men to do anything he would not be willing to do himself. All of these are admirable and desirable qualities but they do not excuse the fact that the supervisor is doing the job himself because he could not get the men who were assigned the task to do it themselves.

Leadership

There are basically two types of leaders, and both types can be equally effective over a short period of time. A supervisor can lead by fear or he can lead by example and motivation. It can be said then that a leader is one who through persuasion or coercion can cause others to do his bidding. The fear referred to is not that of bodily harm but rather the fear of losing one's job, the fear of disciplinary action such as suspension or reprimand, or the fear of receiving a poor efficiency rating affecting potential for future advancements.

Although the use of such disciplinary authority is a valid motivating factor, it quickly loses its effectiveness when it is the only motivating tool being used. Motivational factors will be discussed in more detail in Chapter 3. The most effective leader will be one who by example and motivation achieves the desired results in accomplishing his assigned tasks.

A good supervisor must be an effective leader. An effective leader must possess certain qualities, most of which are learned rather than hereditary. Unfortunately, there are few around who can accurately be described as "a born leader." Consider the following:

1. A leader must be knowledgeable. He must know the job that must be done and the best or the prescribed method for accomplishing the job. He must know the personnel he supervises. It is impossible to motivate personnel without knowing their needs, desires, goals, problems and limitations.
2. A leader must be fair. He must treat all personnel equally in the assignment of work load and duty station, the application of disciplinary action, and in making recommendations for merit awards or promotions.
3. A leader must be firm. He must insure that his instructions are carried out while remaining receptive to new ideas or techniques which might make the job easier or produce better results.
4. A leader must accept responsibility. There is often a fine line between delegating responsibility and passing the buck for failures. If you would take credit for the successful completion of a task, you must also take the responsibility for those that are not accomplished or which have turned out badly.
5. A leader must be tactful. He should praise excellence in public and reprimand poor performance in private. He should avoid losing control of his emotions, yet be able to show enthusiasm for a job well done.
6. A leader must be a capable teacher. He must insure that all instructions, written or oral, are clear and convey to the personnel

the meaning which he intended. He must be able to demonstrate the most effective ways to accomplish given tasks. He must be able to teach the skills required for the accomplishment of the tasks he will assign to his personnel.

7. A leader must take the initiative. He must make decisions when they are required. Nothing is more frustrating than an indecisive person in a position of authority.

8. A leader must persevere. A poor performance should always result in a lesson learned and a resolution to do a better job the next time.

9. A leader must be confident. He must believe in his ability to get the job done right. He must also be able to instill that confidence in the personnel he supervises.

10. A leader must have learned the importance of order and organization. He must have learned to take instruction before he can expect others to follow his lead.

Employee and Management Relations

The new supervisor may have a problem in making the transition from worker to the person responsible for the accomplishment of the work. One of the most prevalent stumbling blocks in this process is the necessary change in attitude and thought. A supervisor is entering the world of management which will require that he begin to think management thoughts and adopt a management attitude.

Instead of just being concerned with the excellence of his individual performance, the supervisor must be concerned with the total performance of the employees under his control. He must consider the effect of his actions and decisions on the effectiveness of the programs he supervises and the business as a whole. He must begin to consider profitability and look for ways to do the work more efficiently to reduce the cost of accomplishing a given task or a series of tasks.

The supervisor must realize that he is the representative of the employees when dealing with management and that he is the representative of management when dealing with the employees. He will find, in most cases, that what is good for the company is usually good for the employees. He must realize that he is not a labor representative but more of a coordinator and communicator. If the efficiency and profitability of the company's operation require manpower cuts, he must be prepared to make them. However, if he firmly believes that proposed action will result in a worsening of the company's condition, he must be prepared to state his convictions and back them up with facts. Once a decision has been reached, he must resolve to follow and support that position, even if he disagrees, unless he is prepared to submit his resignation.

THE SECURITY GUARD FORCE SUPERVISOR

The example given earlier in this chapter illustrated the role of a supervisor as a relatively simple task and pointed out some of the problems that might be encountered. We would like to compare that example with the task facing the supervisor of a security guard force. In the previous example the supervisor had his entire work force under his immediate control and supervision for whatever period of time it took to accomplish the task.

In most departmentalized work forces the supervisor will have his work force under his immediate supervision. There are, of course, exceptions and the security department is one of them. Personnel of the security department are assigned by shifts and by duties to specific posts or patrol areas. They will be found in the gate houses, controlling restricted areas, monitoring alarms and closed circuit television, working in the pass office, providing escort services within and without the protected property, and on patrol routes.

Although there may be two men assigned to a gate or a patrol route, for the most part they will be working independent of each other, and the time available to the supervisor to spend at any given location is limited. He certainly has many other duties to occupy his time. Some of these are assigning and scheduling guard personnel to their post duties, time keeping, preparing post orders and instructions, disseminating changes in orders and temporary orders, developing and scheduling training sessions, conducting training sessions, conducting inspections of all personnel on all shifts, attending meetings and giving briefings to management as well as employees, reviewing daily guard reports, and coordinating necessary follow-up actions, to name a few.

All of these factors make the job of security supervisor much more difficult than supervisors of most other types of employees. The security force must be motivated to perform with a minimum of personal supervision, to use their initiative, to use sound judgment, to follow orders, and to be confident that if they ask for assistance it will be forthcoming without demeaning their ability or without resulting in unjustified disciplinary action.

It is the author's opinion that the position of guard force supervisor is one of the most demanding and difficult positions in the industry. Anyone who is presently successfully filling this position or who is promoted into this position should realize that his contribution to the success of the loss prevention effort is of the utmost importance.

The guard force is the single most important factor contributing to the success of a good security system which is the basis of a risk management program. Given this premise, the importance of the person responsible for the successful accomplishment of the security force mission—the security supervisor—is obvious.

QUESTIONS FOR DISCUSSION

1. Identify the levels of supervision that may be found in a guard force.
2. Name some of the duties required of a supervisor.
3. What are the major differences between the two types of leaders identified in the text?
4. How many "born leaders" have you encountered in your lifetime?
5. What type of specific knowledge must a supervisor possess to be a good leader?
6. Should the supervisor take responsibility for the failure of those he supervises? If so, why?
7. Would it be more effective for the loss prevention manager to make all decisions and not delegate responsibility or authority to his supervisor?
8. Discuss the nature of problems which might be faced by a new supervisor.
9. Should the supervisor stand up for his personnel despite the ramifications his stand might have on the organization that employs him?
10. What factors make the job of the guard force supervisor more difficult than that faced by production line, warehouse or other supervisors?

MAJOR CONSIDERATIONS

1. The ability to lead is an essential quality for a supervisor. Leadership traits are largely learned qualities. Very few, if any, leaders are born to that distinction.
2. A leader must be knowledgeable, fair, firm, tactful, decisive and confident. He must show initiative, accept responsibility, be persevering, be capable of teaching, and know the importance of order and organization.
3. The supervisor must learn to think like a manager but cannot lose touch with the personnel he supervises.

Chapter 3

SELECTION AND
MOTIVATION OF PERSONNEL

The day of the night watchman who is too old, medically retired or other-wise incapable of productive work, watching the premises of a business establishment during the hours of darkness, is over. Please do not mis-understand; we recognize that many persons of advanced years are physically and mentally alert and capable of utilizing their experience to the benefit of society.

It should be recognized, however, that there is no longer an occupation known as "night watchman." Unfortunately, the public image of the security officer is frequently that of an individual who has no more responsi-bility or capability than was attributed to the night watchman. In former years the need for security was much less pronounced, and often the position of night watchman was established to provide a means to keep someone on the payroll who was unable to perform in any other capacity due to the infirmity of age, sickness or accident. This concept is now no more valid than other misconceptions of the security profession.

Historically, many businesses believed the cost of security to be an unnecessary outlay of funds for which they received no financial return. The same belief was once held in obtaining adequate insurance coverage. Bank-ruptcy courts and the record of business failures attest to the fallacy of such beliefs. It is now recognized that inadequate security measures, as inadequate insurance coverage, is tantamount to sending the company treasurer to the casinos of Las Vegas to recoup business losses. It amounts to a calculated gamble of a company's assets.

It may be a sad commentary on today's society, but cities, counties, states and the federal government are no longer financially capable of providing security for the citizens of our country and their property. As a result, a new profession has been born.

The men and women who are pioneering this profession are being called upon to protect billions of dollars in business assets from loss from such sources as fires, vandalism, pilferage, major theft, industrial espionage, embezzlement, bombings and piracy. They are providing personal protection to corporate executives, their families and homes, and they must provide assistance to management in insuring a safe place for the worker to earn his livelihood.

SELECTION OF PERSONNEL

It is quite evident from the foregoing that before we can discuss the nature and degree of training required to prepare for this profession, we must first insure that the personnel selected to perform these tasks are physically, mentally and morally qualified.

It is no longer sufficient to ascertain if the potential applicant has ever been caught or convicted of committing a dishonest act. A thorough background investigation is required, including scholastic achievement; standing in the community (a major endeavor in today's mobile society); stability as shown from previous work records; reputation for honesty, sobriety and morality; ability to manage personal finances; and physical condition.

The standards which must be required of men and women entering this profession may seem unrealistically stringent, but are they? Security professionals are called upon to provide protection for our nation's total assets, and they are responsible for the protection of the lives of its productive work force. Their importance in today's society is as great as our law enforcement agencies and our armed forces.

As a nation, we have rightfully placed great emphasis on the protection of our land, property and individual freedoms from the designs of foreign aggressors. We now realize that this protection will avail us little if we allow our nation's economy to be subverted and destroyed by internal thievery, violence and destruction.

Thus, personnel selected to provide this protection must measure up to the tasks they are called upon to perform. There can be no compromise on their honesty, integrity or morality. The degree of education required will have to be measured by the degree of responsibility they, as individuals, will be required to assume. The minimum, however, should be no less than high school graduation.

They must be sufficiently mature so that having received the required training, they will be capable of sound judgment when called upon to make decisions affecting the protection of life or property. Such decisions will at times be required under great stress arising from emergency situations. The security professional must be fully aware that any lapse in his performance will probably result in the loss of life and/or property for which he is responsible.

Recruitment

Ideally, the security supervisor will want to recruit and employ the cream of the crop of the available work force. Realistically, he will be competing, in the labor market, with other businesses which will often be the higher bidder for the best personnel. It will, therefore, be necessary to emphasize the benefits to be derived from a career in the security profession. Some of the benefits which may be cited are steady, non-seasonal employment; opportunities for advancement into supervisory and management positions for those who are willing to work, learn and be available when the opportunity arises; the opportunity to learn and master a trade; and joining a profession which is somewhat recession-proof.

Recruiting efforts must be aimed at supplying the employment interviewer with the widest possible selection of potential employees from which to choose. Newspaper want ads, employment services and agencies, colleges and universities, military installations, job fairs, fraternal organizations, veterans groups, ethnic organizations and publications, senior citizen groups, and current employees are all excellent sources of potential applicants.

Qualifications

The goal, as in all personnel selection, is to find the best person for a specific job opening. The interviewer should know the job description of each position for which there is a vacancy and conduct the screening interview with an eye towards identifying those applicants who are qualified for the position and making the selection of the best qualified of those applicants.

For example, the position of gate officer will require less manual dexterity and stamina then that of a patrol officer or perhaps an airport check point officer. The gate officer and the airport officer positions would require personnel who relate well with the public and who present a good public relations image. These qualities would not be as essential for the patrol officer. The gate officer and patrol officer should have good writing skills to enable them to submit understandable reports of their activities. At the airport check point, most reports will be written by the shift or check point supervisor.

Naturally, it would be preferable to have all the selected personnel possess the skills necessary to allow them to fill any of the vacant positions. Some of the skills can be learned while others will depend largely upon the applicant's physical make-up, experience, personality and education. If the applicants possess the basic skills called for and have the capacity to learn the other necessary skills, they should be considered as one of the qualified applicants.

The Application

Prior to the pre-employment interview, each applicant should be asked to complete a basic employment application form. This form should at least include the following information:

1. First, middle and last name.
2. Maiden name, nick names or aliases.
3. Current address and telephone number.
4. Social security number.
5. Level of educational achievement, listing the names of schools, locations and dates attended.
6. Previous residences for the past ten years, indicating full address, dates and length of stay.
7. Military service history including information on Vietnam service or service-connected disabilities.
8. Employment history for the previous ten years requiring the applicant to account for all periods of employment or unemployment for the applicable time period. History should include:
    A. Inclusive dates of employment.
    B. Complete name of previous employer.
    C. Address of previous employer.
    D. Job titles and duties.
    E. Immediate supervisor.
    F. Starting and ending wage.
    G. Reason for leaving.
9. Three personal references who are not previous employers or related to the applicant.
10. Statement of applicant that all information provided is true and correct to the best of his knowledge.
11. Signature of applicant.
12. Date of application.

The applicant should be required to complete the application in the presence of the employer's representative to display his ability to understand and properly complete the application form.

Background Investigation

Following the completion of the application form the applicant should be requested to read and sign an information release form authorizing the employer to conduct a background investigation to verify the information provided on the application.

It is good employment procedure to telephonically check a recent previous employer of the applicant in the presence of the applicant. This will make the applicant aware that a background investigation will be made and will provide the applicant with an opportunity to correct or change any incomplete or incorrect information he may have given on the application.

The Interview

The employment interviewer will be given the application, information release, and results of the telephone check of the previous employer to review and to use in the conduct of his interview of the applicant.

Prior to opening the interview, the applicant should be asked to produce a driver's license, non-driver identification, social security card or other appropriate form of identification to verify that the applicant is, in fact, who he claims to be.

Suggestions for the interviewer are provided at the end of this chapter. They can be used by the employment interviewer as a guide in conducting the pre-employment interview.

In many cases the employment interview will be conducted by the security supervisor. If this is not the case, the person conducting the interview should select two or three of the most promising candidates for the vacant position and refer them to the security supervisor for his selection. It is recommended that the background investigation of all applicants considered be complete prior to a final selection. Operational commitments and necessities do not always allow the security supervisor to delay his decision for the time required to gather this information. Should this be the case, the background investigation should be completed as soon as is practical.

MOTIVATION OF PERSONNEL

Guard Force Personnel

Having made the selection, the supervisor must return his attention to one of the most important, challenging and difficult areas that his position demands—the motivation of the guard force he supervises.

Motivation is a never ending responsibility of supervision. Employees must be motivated to continue their employment, to perform their tasks to the best of their ability, to maintain their personal grooming at acceptable levels, to absorb the training required to permit them to perform as intended, and to do the tedious tasks with the same degree of efficiency as they do the more interesting assignments.

Anyone who tells you that motivating a guard force is easy has never had to inspire a patrol officer to approach the third tour through an idle industrial facility with the same dedication and enthusiasm as the first.

The ways in which the guard force is motivated will largely depend upon the type of motivation the supervisor is attempting to achieve. The following are some guidelines that have been proven effective and which can be added to, revised or interchanged depending upon the desired results and the individuals involved.

Money.   This is believed by some to be the ultimate motivator. It can be used in the form of salary and fringe benefits to attract and retain personnel or as incentive pay, cash awards for exceptional performance, or bonuses.

Recognition.   Letters of commendation given for a job well done, news-letters detailing exceptional performance, and establishment of awards to recognize outstanding personnel such as guard of the month, quarter or year.

Rewards.   Promotion from within whenever possible.

Inspections.   Letting the guard force know that the supervisor is aware and concerned with the job being done and that he expects their best performance on each assignment or tour of duty.

Administration.   Ensuring that the guard force is scheduled on a predictable basis and that their pay is both timely and correct.

Personal Interest.   Letting personnel know that their supervisor is interested in their welfare and available to them for discussion of problems concerning their personal life as well as those which are job related.

Changes of Assignments.   Rotate guard personnel where possible to avoid boredom setting in due to the performance of the same tasks each working day. This provides an added benefit of having a guard force that is cross trained and provides the supervisor more flexibility in scheduling, personnel changes and promotions.

Education. An employee who is well versed in his duties and confident in his ability will realize more job satisfaction, is more likely to remain on the job longer, and will perform with greater efficiency than a poorly trained employee.

The Supervisor

The most difficult of all types of motivation for the supervisor is to maintain his self-motivation in the face of the daily frustration of the job.

You may ask, what frustration? Our selection program is flawless, we pay competitive wages with good fringe benefits, our training program is one of the best in the industry, our inspection and audit program insure that our program works. What frustrations could there be?

Guard force supervisors are in a people-oriented business. People are unpredictable. They have individual problems affecting their stability and job performance. Dealing with these problems can be as frustrating as it can be rewarding.

Consider the new employee hired last month to fill a critical position on a guard force. The supervisor took great care to insure the thoroughness of this officer's initial training and job orientation. He is now confident in the officer's ability to perform the assigned tasks. The officer advises the supervisor that a personal emergency has arisen requiring his personal attention and that he is regretfully terminating his services. The supervisor is now back to square one. That is frustration!

There will be times when for every problem a supervisor is able to resolve there will be three which elude him. The supervisor's sense of achievement needs to be focused on the positive aspects of his work. He should realize that in human relations you don't win them all and take pride in the frustrations overcome and the problems which have been successfully solved.

Supervising a guard force is one of the most demanding and difficult jobs in the industry. The chapters that follow should help make that job a little easier.

SUGGESTIONS FOR THE INTERVIEWER

First impressions are not always valid but, if not contradicted, they can tell a great deal about the person to be interviewed. Did he come to apply for this position in a clean, well-groomed condition or is his hair mussed and unkempt, clothing soiled and shoes unshined? A first impression might be that he was not sufficiently interested in the job or perhaps this is his normal state of grooming.

The applicant might volunteer that he came to apply directly from a particularly dirty job and did not expect a personal interview until later. If no explanation is forthcoming, the interviewer should pose a question concerning the appearance of the applicant. Questions may be diplomatically posed, such as asking the applicant if he just got off work, or they may be direct, asking the applicant if he normally goes on job interviews in a state of disarray. Most interviewers will find some middle ground which is comfortable for them without being embarrassing to the applicant.

The initial employment interview is extremely important both to the applicant and the employer. The applicant, if truly interested in the job, will probably be nervous. The interviewer should try to establish rapport with the applicant and put him as much at ease as possible. Use his first name, after determining that the applicant goes by that name rather than a nickname or middle name. The completed application will be helpful here. If a nickname is listed, ask the applicant if he prefers to be addressed by the given name or the nickname.

An attempt may be made to identify with the applicant's birthplace, schooling, former occupation, military experience or some other area of his background. If nothing else, chat about the weather, a recent sporting event or current events in the news.

Review the application, item by item, with the applicant. Be aware of the applicant's response and reaction to questions. For instance, ask the applicant to recite his social security number. Although not everyone has memorized this number, most persons who have been in the work force for any period of time will know their social security number. It has happened that individuals using a brother's, friend's or stolen identification have successfully gained employment under a fictitious or false identity.

While reviewing previous employment history, particularly relating to job responsibility, salary and reason for leaving categories, be alert for any hesitancy or nervousness on the part of the applicant. It could be a signal that all the information is not factual and a more thorough check of these areas is needed.

The employment history will also tell a great deal about the stability of the applicant. If the applicant has jumped from job to job and place to place, this is a strong indication that he will not remain in your employ for any appreciable period of time and the money expended in training and orientation will most probably be wasted. If the applicant appears to be particularly capable and well suited for the position, he should be questioned concerning his reasons for the rather short-term employment and his intentions for remaining in your employ should he be selected. This inquiry should include asking the applicant if he has made application to other potential employers and what his intentions will be if he is offered both positions.

Personal history can also aid the interviewer in making a determination as to the probable stability of the applicant. Is he married or single? Is he the breadwinner supporting a family? Is he a property owner? What is the length of his residency in the area? This is not to imply that the single, solely self-supporting applicant who lives in a mobile home should be ruled out, as he may be the best applicant. These points are, however, valid indicators of potential job stability.

Throughout the interview the personality of the applicant should be continually assessed with the determining factors being how he will relate with other personnel on the job, how he will relate with other employees outside the guard force, if he seems to be open to suggestions or criticism, and how he will probably respond to supervision.

The two remaining determining factors which should be decided by the interviewer are the best placement for the applicant and what his future potential will be. Given two equally qualified applicants for a position where one would be at the top of his capabilities while the other had the potential to progress into more responsible positions, the choice should logically be the applicant possessing the future potential for advancement. This individual will bring a higher degree of proficiency to the job and will be more easily motivated to maintain a higher performance level due to his ability and desire to advance within the organization.

QUESTIONS FOR DISCUSSION

1. What factor has been most responsible for the growth and increasing importance of the security profession?
2. Security officers are employed to protect property. What are some other services that security officers are called upon to perform?
3. Are background investigations of security officer applicants necessary if you have already determined there is no criminal record?
4. Should physical fitness and agility tests be required as prerequisites to employment?
5. Are businesses willing to spend the money necessary to provide an adequate protection program?
6. What is the public's image of the security officer?
7. Does the presence of a security officer on the protected property provide adequate protection?

8. How important is educational achievement in assigning guard force personnel?
9. What are the responsibilities of the security profession today?
10. Does a company really have to spend money on selection and background investigations in order to properly staff their security force?

MAJOR CONSIDERATIONS

1. Use of proper selection techniques to insure the employment of the best qualified applicant.
2. Selecting the motivators most appropriate for the guard force to insure stability of employment, acceptable performance levels, and a consequently successful guard force operation.
3. Understanding that in all human relations there will be defeats as well as victories, taking satisfaction in the victories and learning from the defeats.

Chapter 4

ORGANIZING THE GUARD FORCE

SECURITY CONSULTANT ANALYSIS

Before attempting any organization or reorganization of the security guard force, the services of a security consultant should be obtained to determine the present extent of the security system and to recommend changes in the system which will be designed to provide the degree of protection which the facility requires, in the most economical manner.

In case of a facility that is planned or being enlarged, the time to bring in the consultant is immediately after the decision has been made to construct the facility and before plans, drawings or blueprints of the layout are finalized. Building security into a facility is much more economical than adding it after construction is complete. Most modern structures, due to artificial environmental systems, are already much more secure than older structures. They have fewer entrances and normally have sealed windows rather than those that require locking devices and alarm systems.

The entire system must be effectively implemented with the guard force being the catalyst and the adhesion which binds the system together and makes it effective. A guard force cannot be organized without reviewing the other factors affecting a viable security protection program.

This review must take into consideration all of the following factors:

1. The size of the area that is to be protected.
2. The type of physical barriers that are or will be used to enclose the perimeter of the facility.
3. The number of gates that will be needed to breach the perimeter barrier and the hours of operation of each.
4. The area within the perimeter on which there will be no structures in the foreseeable future, how the area is to be used, and how it will be organized.

5. The number of structures within the protected area, the purpose or function of each structure, and the degree of security that will be required for each structure or for individual operations carried out within each structure.

6. The number of personnel to be employed at the facility broken down by shift, department, salaried or non-salaried, and rate of employee turn over.

7. The number and locations of restricted access areas in use or that will be required, as well as the degree of security protection required for each area.

8. The type and effectiveness of the identification system in use, or planned, to control movement of personnel entering, within or exiting the facility.

9. The type, extent and effectiveness of protective lighting that is planned or provided.

10. The location of employee and visitor parking areas, how they are controlled, and the system used to authorize entry of private vehicles into the protected area.

11. The location of locking devices that are or will be used and the effectiveness of the key control system.

12. The location of shipping/receiving operations, access routes and control measures for company and/or business vehicles permitted to enter the property.

13. The location of installed locking devices and their effectiveness in controlling movement and access.

14. The type and effectiveness of intrusion alarms, where they are located and how they are being monitored.

15. The type of patrol supervision system used or that will be used, its effectiveness and the time required to adequately patrol the area.

16. The location of closed circuit television cameras, their area of coverage and effectiveness.

17. The location and effectiveness of the closed circuit television monitoring system.

18. The type, location and effectiveness of the fire alarm warning system, location of monitors and their effectiveness.

19. The placement, type and effectiveness of protective fire equipment.

20. The effectiveness, location and type of any special access control systems in use or planned.

21. The availability of trained paramedical or medical personnel.

22. The location, suitability and extent of first aid or emergency medical equipment.

23. The type and effectiveness of the communication system in use or that will be made available to the security guard force.

24. The crime rate in the area where the facility is located or where it will be located.

25. The response time of outside assistance which may be required, i.e., police, fire and medical.

26. The type and availability of vehicles that will be provided to the security force.

27. The extent of screening and background investigations that are being or will be conducted on prospective employees.

28. The internal controls that are in effect or planned for the following operations:

    A.   Hiring policies and procedures.
    B.   Payroll procedures and verification.
    C.   Inventory control.
    D.   Shipping/receiving operations.
    E.   Proprietary information.
    F.   Data processing operations.
    G.   Company store operations.
    H.   Credit unions.
    I.   Employee cafeterias.
    J.   Cash sales receipts.
    K.   Employee locker rooms.
    L.   Personnel identification system.
    M. Vehicle identification system.
    N.   Key control procedures.
    O.   Tool, supplies and equipment issues.
    P.   Package control procedures.
    Q.   Mail rooms.
    R.   Expense reporting and payment.
    S.   Company owned/leased vehicle maintenance.
    T.   By-product or salvage sales.

29. The method used to conduct and control trash removal from the property.

30. The effectiveness of implemented or planned safety programs.

31. Any special policies or procedures required for the protection of company executives.

32. The effectiveness of the emergency plan, evacuation routes, evacuation procedures, staging areas, and extent of training being provided or planned for employees and for personnel who have been designated as members of emergency response teams.

33. The number and types of visitors who can be expected to be entering the protected area and the method used or planned to control their entry and movement within the facility.
34. The type of product(s) or service(s) that will be manufactured or offered at the facility to be protected.
35. The extent and projected date of any planned or considered future construction and the proposed location of such construction.
36. Any projects that are currently under construction, type of project, number of construction personnel, type and amount of equipment being used, and the expected completion date.
37. Existing union agreements or proposed agreements that might limit security prerogatives in dealing with employees.
38. Labor laws in the jurisdiction where the facility is located which affect minimum wages, overtime pay requirements, relief and meal breaks or other factors bearing on the ultimate cost of employing a guard force.
39. What budgetary allowances have been authorized or proposed for the employing, organizing and equipping of the security guard force.

Without a personal analysis of all these factors, a security consultant could not begin to give advice on the type and extent of the security force which would be required to provide adequate security for a facility. Some basic information which should be of assistance in organizing and equipping a guard force follows.

PERSONNEL REQUIREMENTS

Supervisory Personnel

The organization of any guard force should begin with the selection and employment of a risk manager or security supervisor. Recruitment efforts should be directed toward finding a manager who is more than a uniformed guard supervisor. The duties which will be assigned to this individual are extensive and require a high degree of tact; require professional knowledge in loss prevention, risk management, safety, fire prevention and control, access control, and emergency situations; and require a leader and motivator of personnel.

Those skills which set apart the risk manager or loss prevention specialist from their predecessor, the security manager, will be elaborated in Chapter 6. If an organization is successful in employing a capable loss prevention manager, the rest will be simple. The staffing requirements, duties, pay scales and other budgetary and operational considerations can be worked out by the new manager working with the consultant or guided by his report.

However, for those businesses that, out of necessity, have delegated responsibility for security to the personnel manager, chief engineer or other staff member as an additional duty, the following basic organizational plan can be used.

A guard force of any size should also have a uniformed supervisor to handle the day-to-day functions of the guard force such as scheduling, assigning, refresher and on-the-job training, payroll preparation, issuing and replacement of uniforms, and daily supervision. The person selected for this position should be given the rank of captain or, at least, lieutenant.

The manager or uniformed supervisor should be provided some administrative assistance in the form of at least one secretary. A large force might require two or more administrative positions to provide typing and filing assistance. It is good practice to staff the pass and identification section with a member of the guard force or security department.

Each duty shift should have a uniformed supervisor to inspect, schedule, supervise and train personnel assigned to his shift and to insure that a responsible person is always available to make necessary decisions. These supervisors should be given the rank of lieutenant or, at least, sergeant. Although it is not recommended, shift supervisors can be used in a patrol capacity or to provide meal and break relief for other members of the force.

For those facilities that ideally employ closed circuit television systems, proprietary intrusion and fire alarm systems, two-way communication with patrols, and other valuable adjuncts to a good security system, it will require one person per shift to maintain the shift blotter and monitor the installed systems. It is preferable to have the shift supervisor free to roam the area and conduct inspections of personnel, but it may be necessary, for budgetary reasons, to use the shift supervisor in this position.

Gate and Patrol Personnel

Each gate that breaches the perimeter of a protected facility must be manned by a minimum of one guard at all times the gate is open. Gates that have a continuous flow of incoming and outgoing traffic will generally require two officers to properly handle the traffic. At gates where heavy traffic is experienced only during periods of shift changes, a patrolling or reserve officer may be dispatched to provide assistance to the guard assigned to the gate.

The number of officers to be assigned to patrol duties will be deter-mined by the extent of the area to be covered, whether patrols are made on foot or by vehicle, the availability and use of closed circuit television, alarm systems or other elements which might reduce the need for manpower. As a rule of thumb, a patrol route should take no longer than forty-five minutes per tour to adequately patrol a given area. Thus, if the area to be covered required three hours to adequately patrol, four patrolmen per shift would be needed.

Closed circuit television and alarm systems can only be as effective as the response provided to investigate suspected activity seen on the monitor or reported by the alarm system. Unless patrol routes are set up in such a way as to permit the officers on patrol to respond to any situation in a reasonable time, provisions must be made for relief guards, in reserve, to answer such alarms or observances. When these officers were relieving patrol or gate officers, the officers on break would be required to respond to the alert.

Guard Force Model

Given a facility that contained three gates open twenty-four hours per day, seven days per week, required three hours to properly patrol, and had installed and in use closed circuit television and proprietary intrusion and fire alarms, the structure of the guard force would approximate the strength listed below.

It should be noted that in preparing a budget and cost analysis for the operation of a guard force, the risk manager and his administrative staff, including the pass and I.D. officer, should have their time computed on a pro rata basis, charging to guard operations only that portion spent in support of that function.

TITLE	NO. PER SHIFT	TOTAL
Risk manager	N/A	1
Captain	N/A	1
Secretary	N/A	1
Clerk/typist	N/A	1
Pass & I.D. officer	N/A	1
Shift supervisor	1	4 + 8 hours
Dispatch/monitor	1	4 + 8 hours
Gate guards	3	12 + 24 hours
Patrol officers	4	16 + 32 hours
Reserve officers	2	8 + 16 hours
	Total	51 + 8 hours

These totals can be affected in many ways, including the hours the plant or facility is in operation, the days per week it is in operation, the number of employees assigned to each shift, and the hours of operation of the shipping/receiving departments, to name a few.

Gate houses should be provided at all gate locations which are open on a daily basis. These houses should be of sturdy construction, be wired for electricity and for telephones, provide heating and cooling devices of sufficient capacity to handle temperature extremes, and provide the guard with the ability to observe any event taking place in a 360-degree radius from the guard house (Figures 4-1 and 4-2). If reliefs are not provided, toilet facilities should be considered.

COMMUNICATION DEVICES

It will be necessary to provide whatever communication devices are required to allow the security force to operate effectively. Patrol officers, reserve officers and supervisors will, at times, be needed on a minute's notice to respond to an accident, incident or alarm. Unless there is a public address system which reaches every point within the facility, security personnel not on static post should be equipped with some method of being contacted.

Probably the best method of communication is to provide portable two-way handie-talkies connecting roving members of the security force with the dispatcher or guard office (Figure 4-3). If this method is used, extra batteries must be obtained and battery chargers provided. A policy should be

Figure 4-1.   Gate house at entrance to protected mobile home park. Photo by author.

Figure 4-2. Gate house at entrance to protected residential area. Photo by author.

established detailing the frequency of battery changes and the length of time required to recharge batteries. The best system is to provide each shift with their own, distinctively color-coded battery packs which they are required to use during their shift and are responsible for putting in the battery chargers at shift's end. It is a good idea to carry this a step further by numbering the radios and using the corresponding number for the battery packs to be used with each radio.

In facilities where telephones are available in close proximity to any point along the established patrol routes, telephone beepers might be carried by the patrolling guard. Although this is not as efficient as two-way communications, it is a good alternative. Another method of maintaining communication is by requiring roving personnel to call in to the dispatcher periodically. This method is not a good one, but if other alternatives are economically impossible, it is preferable to having roving personnel out of touch for long periods of time.

It is the author's recommendation that regardless of how other members of the security guard force are equipped, the supervisor should have the capability of two-way communications with the dispatcher at all times.

Even with a professional risk manager on staff, it will still pay to contract with a security consultant to conduct a survey, and provide recommendations on how best to secure the facility and to organize or reorganize the guard force. The outside consultant will see the system a little differently than the man in charge and will be able to make recommendations or

Figure 4-3.   Hand-held portable radio and battery charger. Courtesy of Wilson Electronics.

comments without being concerned with any possible adverse reaction by management. Most risk managers will welcome this assistance since in most cases the consultant report will reinforce the manager's own recommendations and suggestions.

QUESTIONS FOR DISCUSSION

1. What is the purpose of having a security consultant review the security procedures in effect at a facility prior to organizing the guard force?
2. What is the first step to be taken in organizing a guard force?
3. How does the scope and extent of area within the protected facility affect the staffing of a guard force?
4. How important is an effective communications capability to the overall efficiency of the guard force?
5. At what point in the development of a new facility should a security consultant be employed?
6. How does the response time of outside agencies affect the composition and strength of a guard force?
7. Discuss the importance of establishing effective internal controls as they relate to the overall efficiency of the facility's security system.
8. What effect, if any, does the installation of closed circuit television have on the strength of the guard force?
9. What is the purpose of employing reserve or relief officers as members of the guard force?
10. What role do you see the security guard force fulfilling in the overall security/loss prevention system at the facility where you are employed?

MAJOR CONSIDERATIONS

1. The guard force is the catalyst which activates and unifies the security system.
2. Organization of the guard force cannot begin until a total review of existing conditions or planned implementations has been made.
3. The guard force provides response to alarms, controls personnel and vehicle movement, enhances employee safety and eliminates or reduces hazards.

Chapter 5

SECURITY HAZARDS

From the definition of security given earlier, it is obvious that if you have security, you must have reduced or eliminated all hazards. Where hazards exist, you are still working toward the attainment of a viable security system or your system has been breached.

It would seem that a good definition of a security hazard would be any threat or condition which, if not neutralized or controlled, would breach the protective screen that constitutes the security of the protected property, area or individual. Needless to say, such a breach could cause the compromise or loss of company secrets, loss or destruction of company property, and/or injury or death of personnel. Any of these factors would ultimately result in a loss of profitability for the business.

Professional security supervisors or officers must be knowledgeable of the types and nature of hazards they are or may be facing in order that they be constantly controlled, where possible, and plans formulated to deal with those hazards that are beyond their control. In this chapter the major hazards faced will be identified, the duties of security personnel from the risk manager to the security officer will be defined, and some techniques which may be employed to identify, reduce or eliminate the hazards will be provided.

Security hazards may be divided into two distinct groups: those hazards brought about by the acts or omissions of human beings are referred to as **human hazards** and those hazards over which we have no control, often referred to as works of nature or acts of God, are called **natural hazards.**

Through the use of good planning and implementation of a viable security program, human hazards can be eliminated or reduced to an acceptable level. Through careful and thorough planning, beginning with site selection and continuing through the architectural and construction stages of facilities, a good deal of control can be exercised over the effect of natural hazards. The effects can be further minimized by implementing an effective plan for emergency procedures (Chapter 18)

For ease of recognition, the major hazards are listed below in the two groups:

HUMAN HAZARDS	NATURAL HAZARDS
Theft	Floods
Pilferage	Earthquake
Accidents	Violent Winds
Sabotage	Heat
Espionage	Cold
Fires & Explosions	Fires

It may appear that hazards have been duplicated by listing both theft and pilferage. It can be said that all pilferage constitutes theft but the same is not true in reverse.

HUMAN HAZARDS

Pilferage

A pilferer is normally authorized within the protected area and is one who steals in small quantities. Because of the small quantities taken each time, the pilferer has often been thought to be a small problem. However, due to the frequency of pilfering and the number of persons involved, these small amounts soon multiply into huge losses for the protected business.

Theft, by pilferage, is probably the greatest recurring threat to the security system that the guard force will encounter. For this reason, special emphasis should be placed on insuring that the guard force is well versed in the ways and means the pilferer will use in attempting to breach the protective system.

Security officers should be aware that there are two types of pilferers. The *casual pilferer*, who steals as the opportunity arises, will usually keep the item for his own use, give it to a friend, or sell it as soon as possible. The casual pilferer is not a risk taker. That is, he will not steal unless he is confident that the chances of being caught are minimal. The best and most effective deterrent to the operations of the casual pilferer is the use of unannounced lunch box, package or purse inspections. The risk of being caught in an unannounced inspection cannot be justified by the small gain he would realize.

The second type is referred to as the *systematic pilferer*. As the name implies, this person has devised a plan or system to regularly steal from the protected area. His plans are made after the security plan has been implemented; therefore, he has been able to design his plan around the weaknesses in the security system.

Security officers who are faced with a systematic pilferer should be aware that they are dealing with an organized group rather than with one person acting independently. This type of thief surrounds himself with accomplices drawn from his fellow employees, truck drivers who make pick-ups or deliveries to the protected area, other business visitors, or outsiders. His plan will include every phase of his operation from the actual theft, the removal from the property, to the disposal of the stolen goods.

When a systematic pilferer is operating within a protected area, the security screen has been breached and rendered ineffective. Since the pilferer has breached the protective screen and is moving the stolen goods from the property to a point where they can be disposed, he is often able to continue the operation for weeks and months without detection. Detection, when it comes, will usually be the result of losses being discovered by inventory.

It will be necessary for the security supervisor to instill in his guard force the need to constantly be alert for methods that a pilferer might use to defeat the established security system. Both the supervisor and the guard force must continually review the measures being taken and try to devise new methods which make it increasingly more difficult for anyone to breach the security screen.

Many firms, having programmed their inventory control into their computer data banks, feel that they are always on top of the stock on hand. They fail to realize that the computer only records the data that it has been fed. Thus, the amount of inventory received, less the amount used in production or shipped should leave the balance on hand. Without a physical count it cannot be determined if a theft of property has taken place. There is also the possibility of the key punch operator being part of the theft ring and falsifying receipts or shipments to delay discovery of the loss. Although the security force has no immediate control over these methods, they can, as will be pointed out below, discover shortages or irregularities in inventories which will alert them to the probability of the presence of a theft ring.

The systematic pilferer's target will normally be the small, easily transported but expensive items which are easily disposed of in the market place. The guard force, by conducting spot inventories of these items, will often be able to identify areas of loss without waiting for a physical inventory of the entire stock on hand. An easier and more effective means of spotting such losses is establishing visual inspections of such items by the patrolling guards. When cartons are found to have been tampered with or moved and records indicate no shipments have taken place, a method of surveillance can be established on these items, giving the guard force the opportunity to catch the thief in the act of taking the property. The actual taking of the item is usually not difficult. The critical area for the pilferer is the moving of the item through the security screen to a place where it can be sold.

By making the guard force aware of the methods which will probably be used by the systematic pilferer and insuring that they are alertly guarding against them, a giant step will have been taken in preventing such persons from operating. Since the thief relies on poor accountability and inventory procedures to allow the losses to go undetected, it is essential that all such procedures be periodically reviewed. This review is properly the function of the inspection and audit division within the security department.

Once it is known or suspected that a systematic pilferer is operating, the supervisor and other members of the guard force can more thoroughly scrutinize ways the pilferer might use to move or dispose of the stolen property. In doing this they will have to keep in mind any methods which may be available to the pilferer which are unique to their operation.

In most cases the risk manager and other management members of the business will have a good idea of the department or division from which the stolen items are being taken. In these cases an undercover investigator can be hired to fill a slot in the suspected area. He should be processed in the same manner as every other employee and his true identity should be known only by a highly selected few. The investigator will attempt to identify the thief, gain his confidence, and infiltrate the group so that all persons involved in the stealing, transport and disposition of the stolen items may ultimately be identified and apprehended.

When conducting briefings or training sessions the supervisor should periodically review the various methods which might be used by a pilferer to move stolen property through a security screen. The following paragraphs describe methods which have been used and which are generally adaptable to the average protected facility.

Plain View.   In cases where the protected property is a warehouse or dock storage area containing items too numerous for the security force to recognize, a worker will often exchange items of clothing. This is done simply—the thief wears the new clothing, leaving the old clothing he wore to work in its place. He will normally have taken the precaution of scuffing the shoes or putting dirt on the shirt or trousers so they do not appear new.

Clothing.   Nothing could be easier than carrying out small items in the pockets of clothing, wearing items under regular clothing, or taping items to the body. This is particularly effective since security officers may not search a person without having probable cause to believe the individual has stolen property on his person. Even in such cases, in the absence of the power to make an arrest, the law in most jurisdictions requires that a search warrant be obtained prior to conducting a search.

Packages. If the security plan allows personnel to leave the protected area with packages, without control, the pilferer will take advantage of this breach in the system and use this method of moving stolen material through the security screen. Whenever packages are permitted to be taken out of the protected area, a good package control pass system must be used. The person authorized to sign the pass must inspect the contents, seal the package and execute a pass detailing exactly what the package contains. Even when such a system is in effect, security personnel should conduct spot checks of package contents to insure the integrity of the system.

Personal Property. Employees, vendors and visitors will often bring to and carry from the protected areas items of personal property, such as lunch boxes, purses, briefcases, etc., which are capable of holding small items of company property. The best method of preventing the removal of stolen articles inside these containers is by conducting unannounced inspections of these items. It is a good idea to have signs prepared and posted indicating that such items are subject to search upon departure. Provisions should be made for personnel entering the area to check these personal items should they not want to be subjected to the search.

During the inspections it is good policy to have a member of management available to handle any employee found to be in possession of company property. It should be pointed out to the members of the guard force that these are inspections and not searches. Accordingly, items found during such inspections will normally not be admissible in a court of law as evidence of theft. These items may be confiscated for return to the company, and administrative action, as deemed necessary, may be taken against the employee.

Motor Vehicles. Private automobiles which are permitted to be parked within the protected area, without provisions for unannounced inspections, are an open invitation to the pilferer. As with other vehicles permitted to enter the protected area, private automobiles may be used to transport stolen property with or without the knowledge of the owner.

The pilferer, acting in collusion with truck drivers, may overload a truck and secrete stolen items in the cab or on the undercarriage of the truck. In the event the truck driver is unaware of the stolen property he is transporting, the pilferer will usually have a confederate outside the area who will remove the items once the truck has left the protected area. Because these are excellent routes for the pilferer to channel his stolen goods through the protective screen, provisions should be made for conducting inspections of all vehicles departing the protected area.

Openings in the Perimeter Barrier. In cases where the walls of a building form part of the perimeter barrier, the doors and windows in the wall offer the pilferer a good method of disposing of stolen property. Goods may be thrown out for a confederate to pick up or for later recovery by the pilferer himself. Needless to say, any gate that breaches the perimeter barrier should be either locked or controlled at all times. These breaches may be eliminated entirely by using proper physical security measures, as discussed in Chapter 7.

Trash and Garbage Removal.   In facilities where all refuse is not put through a compactor prior to being removed from the protected area, pilferers often send out stolen property with the trash. Installation and properly controlled use of a compactor will effectively eliminate this method. Without compacting equipment, guards should conduct unannounced inspections of the refuse removal operation and inspect the trash being removed either at the loading point or by accompanying the trucks to the dump site.

Rail Cars.   Much the same as motor vehicles, rail cars that enter and depart the protected area provide an excellent means for the pilferer to move the stolen property. Since most rail cars are sealed prior to departure, the pilferer must either be in collusion with railroad employees or must limit his use to the undercarriage or other outside areas of the cars. Guards who are required to open rail gates for train movements should carefully check each rail car prior to its departure.

Miscellaneous.   Guards at facilities having streams which flow through the property should be alert for stolen property being placed in plastic bags or containers and floated outside the protected area. While on patrol throughout the facility, guards should be constantly alert for property which is out of place. The pilferer will remove the items to be stolen from their designated location and wait to see if the loss is discovered before attempting to move them from the property. Pilferers also use a step by step method, that is, moving the stolen property a step at a time until it is near a place where it can be easily removed from the property.

These are but a few of the methods that can be devised by the pilferer to remove property through a protective screen. The supervisor and his guard force must always be on the alert for any weakness or a potential breach in the security of their protected area.

It might help the guard force to better understand the problem of employee theft if they are made aware of how the problem starts and what contributes to its growth. The vast majority of the employees are basically honest and many will not steal even if the opportunity to do so presents itself. Poor security is usually a contributing factor to the beginning of employee theft.

In the beginning the employee might take an item or two for his own use. Possibly a neighbor will ask if the employee can obtain a similar item for him at cost. This procedure can grow through supplying friends, then friends of friends, until the proceeds of the thievery have become a quite modest extra income for the employee. It is human nature that when you have more you tend to spend more. The employee often finds that he must continue to steal, and in quantity, in order to maintain his present living style and be able to pay off his creditors.

If efforts to identify the method used to remove the pilfered items and attempts to infiltrate the operation have failed, all is not lost. You still have identifiable property being fenced or sold. By determining who is selling the property it may be possible to trace the stolen goods back to the pilferer inside the protected area and in the process learn the method which had been used to remove the property through the security screen.

This operation can best be handled by employing an outside investigator to identify the outlet for the stolen goods. An alternate method would be through liaison with the local police department, alerting them to the problem and requesting their assistance in determining the persons responsible for the pilferage.

Accidents

Accidents, whether caused by carelessness, negligence, or total disregard for life and property, can be as great a danger to the protected property and to those working within the area as a major flood, earthquake or other great natural hazard. It was once believed that accidents will happen, but we now know that accidents do not just happen — they are caused.

A well-trained guard can assist in the prevention of accidents by reporting violations of safety procedures such as reckless operation of lifts, trucks and other motorized vehicles; failure to wear prescribed safety equipment; use of fire or smoking material in unauthorized areas; poor house-keeping practices; and any other safety hazards encountered during his tour. Accidents and their prevention will be discussed more thoroughly in Chapter 17, dealing with employee and facility safety procedures.

Espionage

Many people still believe that spies are always foreign agents gathering intelligence information and compromising our national secrets. The story books or movies referred to as spy thrillers always have the hero using ultra-sophisticated equipment, surrounded by beautiful women and driving fast, expensive cars. Since modern technology can make it possible for one firm to out-produce another or to produce a superior product for less money, industrial espionage has become big business.

In today's society, the protection of company secrets from compromise is of major importance. Do not expect to see a spy driving up to a security check-point in a Mercedes Benz, 300 SL loaded to capacity with beautiful girls scantily clad in bikini swim suits. Although this would be a unique approach, the industrial spy will more likely attempt to breach the security screen by using subterfuge. He may be the electrical contractor, telephone repairman or vendor equipment maintenance man. He is not likely to be on the list of persons who have been authorized entrance on a temporary basis. He will know the names of top management and will be a convincing portrayal of an honest man just trying to do the job for which he is being paid.

As a rule of thumb, to complement identification systems and visitor rules, it should be strictly practiced that no unauthorized person will be permitted to enter any area where proprietary information is kept, unless he is constantly accompanied and under the scrutiny of a member of management or the security guard force.

In today's world of sophisticated electronics it is not sufficient to merely exclude the potential industrial spy from access to written or graphic illustrations of proprietary material. It is also necessary to provide adequate security protection to the spoken word.

Board rooms, research and development laboratories, testing areas and any other areas where proprietary information will be discussed must be periodically "swept" to insure that no listening devices have been planted in the area. Care must also be taken to insure that no one introduces such devices into the area on their person.

Board rooms and other areas where proprietary information is discussed should be soundproofed. During such discussions, windows should be covered to preclude eavesdropping from outside the property through the use of sensitive directional microphones.

Preventing the introduction of an industrial spy by means of employment is best achieved by conducting thorough background investigations and, in the case of key personnel, follow-up checks at regular intervals. The use of the polygraph on a periodic basis is also useful in deterring potential industrial spies.

By following these procedures, as long as other security measures are adequate, it should be possible to effectively prevent the industrial spy from achieving his goal.

Sabotage

There is little distinctive difference between damage caused by vandals and that caused by saboteurs, bombs planted for social protest or those planted to injure that particular property, or, for that matter, a fire set intentionally by a pyromaniac or a person bent on causing extensive damage to the protected property. Establishing and maintaining a good security screen with adequate rules and procedures governing entrance, exit and movement within the area will provide the same protection against sabotage or espionage as it does against theft or pilferage.

The breakdown can and often does come in the employment, identification and control of personnel. As stated, the pilferer is generally authorized to be in the area from which he steals. Poor employment policies, lack of proper screening, and lack of a viable identification system often result in the thief, saboteur, arsonist or spy being authorized to be within the protected area.

When these circumstances exist it makes the job of the security department doubly difficult. It will not usually be known that a saboteur is inside the facility until he has done some damage to the property therein. In this case the job is more of an investigative nature rather than preventative. However, unless the perpetrator can be identified, he is likely to do more damage.

As in any investigation, motive and opportunity will have to be established before it is possible to identify the person responsible. To establish opportunity it is necessary to determine the exact time the act of sabotage took place and determine all persons who had access to the area where the act took place at that time. Once it has been established who had the opportunity, a check can begin of all individuals in that category to determine who had a motive for damaging the equipment or facility. Unless the supervisor has had investigative experience or has investigators assigned to the security force, it would be a good idea to call in the police or professional private investigators.

This is not to say that the security guard force should do nothing. Patrolling guards and guards assigned to gates or other strategic locations, by being ever alert and watchful, may often detect the perpetrator of sabotage due to their furtive and suspicious actions.

Alertness is one of the most important attributes of a competent, well-trained, professional security officer.

NATURAL HAZARDS

As outlined before, natural hazards pose a problem that is difficult to deal with. If they are to be controlled to the extent that they will not disrupt operations or result in the loss of property or human life, then a great deal of prior planning must be done.

The planning must begin at the point where construction of a facility is being contemplated. The selection of the building site is of extreme importance. For instance, if the plant were to be built in California, care would have to be taken to insure that it was not located on an earthquake fault line or was not in an area where it might become a victim of a mud slide or forest fire. In the desert, care must be taken to insure that the site is not in an area known for flash floods. In valleys which are downstream of a dam site, the facility should be placed on high ground to guard against the possibility of a future flood condition requiring the opening of flood gates or even the rupture of the dam itself.

The guard force has little control, if any, over the initial planning and is often left out of the preparation of plans for implementing emergency procedures. Whether or not consulted, the security force has great responsibility in the implementation of such plans and has the responsibility to provide input to management concerning necessary changes or improvements which should be made in these plans.

The guard force, due to its patrols within the protected area and continuing training, should be well versed on all aspects of fire, safety, lighting, emergency exit locations and other areas that go into establishing a workable plan for emergency procedures. The degree of involvement of the guard force will be made more apparent in subsequent chapters dealing with these subjects. The security supervisor should require his guard force to keep him up to date on anything that would affect the emergency procedures established for his area of responsibility.

Examples of the types of reports which should be required would be erosion of land in staging areas; poor maintenance of emergency equipment, such as lights, first aid kits and communications; blocking of emergency exit doors, lanes or staging areas; and blockage of flood channels or drains which could make damage more severe during a flood emergency. The supervisor would submit some of these reports to management immediately, while others could wait for scheduled meetings or planning conferences to update and improve the emergency procedures.

Construction is also important in withstanding natural hazards. The facility must be built sufficiently strong to withstand any natural hazard native to the area where it is to be located. These hazards could include floods, hurricanes, earthquakes, tornados, lightning and must even include protection against freezing cold or extreme heat.

In eastern Texas when the Gulf of Mexico annually leaves it banks and floods the adjacent areas, a twenty-one-story building has incorporated flood control gates designed to permit flood waters to fill the basement. Water-tight doors and construction materials prevent the water from going higher in the building with the result being that the building, in effect, floats instead of being buffeted by the flood waters.

Buildings in earthquake-prone Los Angeles and San Francisco are constructed in such a way so as to allow them to sway or bend with the movement of the earth rather than remain rigid and be torn apart by the earthquake. The volcanic eruption of Mount St. Helens, Washington in 1980 rocked the Pacific Northwest with an explosive power equivalent to the detonation of a nuclear device. No one foresaw the occurrence of this disaster, but hastily formulated emergency plans resulted in the saving of many lives which otherwise might have been lost. If, or when, this area is reconstructed, a great deal of prior planning must take place concerning the location and construction of planned buildings, logging camps or commercial ventures.

In conclusion, human hazards can be controlled by eliminating or minimizing the opportunity for their perpetration. Natural hazards can neither be controlled nor minimized, but the effect they have on the facility and production capability can be anticipated and minimized through prior planning.

QUESTIONS FOR DISCUSSION

1. Name the two types of pilferers and explain how they differ.
2. Define human hazards.
3. Give some examples of human hazards.
4. Is it possible to minimize or prevent losses caused by natural hazards?
5. Name some of the conditions which allow a systematic pilferer to operate.
6. Will a good package pass system eliminate the problem of pilferage?
7. What steps can be taken to deter an industrial spy from successfully penetrating the security screen?
8. What actions can be taken to deprive potential pilferers access to the protected property?
9. What can be done to prepare the guard force to effectively handle security hazards?
10. What measures can be instituted which will minimize the threat of sabotage?

MAJOR CONSIDERATIONS

1. Security planning must include safeguards against natural hazards as well as human hazards.
2. Pilferage and theft differ in the thieves' accessibility to the protected area.
3. Recognizing the threats is essential if they are to be eliminated or controlled.

Chapter 6

LOSS PREVENTION

In previous chapters risk management and loss prevention were referred to and it was stated that these titles more accurately describe today's security director or manager. Before getting into some specific examples of loss prevention techniques, it is necessary to define these titles and the additional duties brought about by these changes.

The insurance industry has and continues to identify their specialist in the prevention of loss as a risk or loss prevention manager. In adopting these titles and duties the security director is expanding his traditional role of the prevention of crime, i.e., pilferage, theft, arson, sabotage, espionage and the prevention of loss-producing accidents, to providing his company with a total loss prevention program.

Where the insurance or loss prevention specialist is primarily concerned with risks that are insurable, the security/risk manager must be concerned with all risks and determine the most logical and cost-effective method of reducing, eliminating or dealing with these risks.

LOSS PREVENTION AND PROFIT

A risk can be defined as *anything* that adversely affects the profitability of a business. Loss, as used here, refers to the bottom line of a business financial statement. In this context, all losses equate to reduced profits for the business. Could it be said then that preventing losses should result in a comparable increase in profits? Not necessarily; if it were that simple all businesses could be protected like Fort Knox and automatically become more profitable.

Identifying the point where providing protection to assets turns from profitable to nonprofitable is the task of the risk manager. In order to be effective he must identify all potential areas of loss, determine the probable occurrence and extent of loss in each area, and identify the controls or protection required to prevent or reduce the loss. Once these factors are known, it becomes a matter of simple arithmetic. If the loss, left unchecked, would be less than the cost of implementing controls or protection, then it becomes more profitable for the business to accept the loss.

In most cases it will be possible to reduce or eliminate the loss through the implementation of controls and/or protective devices or services which will result in an effective increase in the profitability of the business. To accomplish this requires a careful balance of protective services and insurance coverage which are constantly reevaluated to insure their validity.

For example, a manufacturer of television sets is losing picture tubes from its warehouse, with an annualized dollar value of $50,000.00. The insurance premium covering losses from the warehouse is $5,000.00 per year. It has been determined that adding a security patrol in the warehouse during the forty hours of weekly operation at an annual cost of $12,500.00 would effectively reduce or eliminate the losses. It seems evident, on the surface, that obtaining the insurance coverage would be the most economical solution.

However, insurance companies are businesses which are also concerned with their bottom line profitability. Their risk manager will identify the problem area and will require, as a condition of insurance, that the risk be reduced. The insurance company has the option of refusing coverage if their conditions are not met or of increasing the premiums to absorb the losses. In either case, the solution for reducing the risk with an increase in profitability would be the employment of the patrolling security officer. Insurance coverage is essential to protect the business from unforeseeable or catastrophic losses. A business that expects its insurance carrier to continue to pick up the tab for predictable and preventable losses is in for a rude awakening. For those corporations which qualify as self-insurers, the cost of the loss is equated directly with the cost of protection.

INTERNAL CONTROLS

The devices, systems and forces which will be discussed throughout the chapters of this book are primarily concerned with preventing losses from criminal acts, accidents and the natural hazards identified earlier in this chapter. There are other loss potential situations which can be predicted and must be identified and protected against. Some of these seem far afield from a security director's duties and responsibilities but are properly the concern of the risk manager.

Take, for example, the multi-national U.S.-based pharmaceutical firm operating a manufacturing plant in South America. Labels for the drugs being produced are printed in Spanish, English and Portuguese, depending on the area in which they are to be marketed. The bilingual employee normally responsible for loading the labeling machine for the day's production is out sick; his assistant, who speaks all three languages, reads only Portuguese. The day's production is in English and the wrong label is used, identifying the product as an entirely different drug than that packaged. The day's production is packaged, labeled, cased and shipped before the mistake is discovered. The loss potential is enormous. If the mistake is discovered in time, the shipment can be recalled but still must be tested, positively identified and relabeled. If the mistake is not discovered in time, the result might be death or serious side effects to the consumers, loss of consumer confidence in the product, and lawsuit upon lawsuit. This type of loss could be prevented by instituting controls that require a mandatory written test for anyone working in a critical position demanding bilingual reading skills and having a supervisor or quality control inspector check each separate production run to insure that the proper label was being used.

Today's risk manager must be knowledgeable in all areas of the business in which he is employed, including production, finance, marketing, warehousing, advertising and legal matters as well as the protective services. His corporate relationship, as depicted in Figure 1-1, has him reporting to the chief operating officer and necessarily interfacing with all other directors and department heads.

A few years ago a study was conducted in a large petrochemical production facility which had lost the market for its primary money-making product. The loss of market had come about due to improved refining processes which permitted refineries to produce the product at a lower cost than could the petro-chemical company. Losses were accepted, during the good years, when little attention was given to expense accounting procedures or controls, using bid procedures to obtain the best price for needed products or repairs, unexplained losses of raw material or of products. It was believed that such controls were not necessary for, after all, plenty of money was coming in and a good profit was being made. Had these controls been in effect, the cost of producing the product would have been less, which might have enabled the company to compete with the refineries. In any event, the company probably could have survived the necessary outlay to convert their facility to other products. Since there were no controls, the firm had to close down and sell out.

Losses in the millions of dollars are taking place in many businesses due to inadequate controls and protective services. The hotel industry has always been a prime example. The losses incurred are written off as a cost of doing business and the cost is passed on to the consumer using these facilities. The casinos of Nevada are victimized by skimming operations, which are the taking of unreported income and diverting it to other businesses or interests usually operated by organized crime. Yet, with the exception of large, well-organized corporations, few of the casinos employ professional risk managers. They employ elaborate surveillance devices and usually well-trained and well-paid guard forces but leave the controls up to state gaming authority agents or outside private agencies with which they contract.

Security Survey

The risk manager will be guided in his decisions and recommendations by the technical services division, which will conduct security/loss prevention surveys of all facilities that come under its area of responsibility. A security survey is a study which analyzes and provides the answers to all the factors listed in Chapter 4, as well as threats posed by natural hazards native to the area. Further, it provides recommended solutions to all factors which are found to be weak or inadequate.

The security supervisor will also find the survey an invaluable tool in making decisions concerning the deployment of his personnel, establishing patrol routes, preparing post orders, training the guard force, and in many other areas of his day-to-day responsibilities. Conversely, the consultants conducting the survey will enlist the aid of the supervisor to provide many of the answers on which they will base their analysis and recommendations.

It may seem to the supervisor that the security screen is impenetrable since it prevents property that has been stolen from being carried out of the protected area by employees and visitors or in vehicles. As will be shown, a firm can still be incurring losses of property that is being removed in a manner that, on the surface, appears to be sanctioned by existing operating procedures.

Recommendations for the establishment of controls to prevent these losses will be made by the consultants in their survey and presented to management by the risk manager. The supervisor and the members of the guard force can assist in this process by knowing the areas where these losses may occur and reporting deficiencies in these areas in their daily reports. Once appropriate controls are in place, they must still be checked and double checked at each level of company operations in order to maintain their viability.

Shipping and Receiving Operations

Members of the guard force can begin by checking controls of incoming shipments of raw materials, components, spare parts and finished products. In

many cases shipments are signed for but not received or perhaps are only partially received. This may not be a problem, but without safeguards it can become
a problem. Some suggestions can be offered here, but without being familiar
with the particular circumstances of an operation, cut and dried solutions
cannot be offered.

A suggestion for correcting possible discrepancies in receiving operations
is to assign someone to periodically spot check incoming shipments to insure
their correctness. This can be done by security officers or supervisory personnel. The same system may be used for double checking shipments leaving the
protected area.

A security bay or area needs to be set aside. Periodically the security
officer, or other designated person, should divert to the security area a shipment which has been receipted for as accurate, where it will be rechecked
against receiving documents. To apply this to shipping operations, once a shipment has been pulled and is ready for loading, it can be diverted to the security
area for a verification check. This should include not only a check of quantity
of items but should also verify that the addresses where the items are to be
delivered are the same as on the shipping documents.

Another area which should be checked is the mail and shipping operations. If labels are not controlled, stolen property can be sent anywhere by
mail or shipping department personnel or even outside personnel who have
access to shipping and mailing labels.

A survey was conducted of a national air freight company's operation
which was losing valuable shipments and incurring large insurance claims. In
checking their system it was found that, due to the nonexistence of controls on
shipping labels, any package in the system could be diverted at any point in the
system. This could easily be accomplished by relabeling the designated package
and sending it to a confederate. This company was actually delivering the
stolen property for the thieves. These thefts were going on completely
undetected, and the cost of putting undercover investigators in each of the ter-
minals around the country was prohibitive.

The adoption of sequential serial numbers on all shipping labels was
recommended, making the labels difficult to counterfeit by printing them on
controlled paper using a color-coded intricate design and imposing absolute
control over the use of shipping documents and labels. This process can be
even more effective by changing color coding of labels without notice. This
control can be used to control package shipments from mail rooms, parcel
post and freight operations. Shipping labels are issued only to authorized personnel and then by serial number. Serial numbers not appearing on shipping
documents must be accounted for at the end of each shift. Labels which have
been mistyped or contained errors must be voided and returned for accountability. The control of labels coupled with spot checks of shipments should
considerably tighten the security of shipping operations.

Personnel

The following paragraphs will provide some ideas of where to look for weaknesses in security systems and controls by listing some of the innovative ways that employers have been and continue to be bilked of property or income that is rightfully theirs.

The theft of time has always been a favorite. This usually starts with an employee leaving the job an hour or two early for a medical appointment or some personal interest and having a friend, his supervisor or sometimes even the security officer clock him out when his regular shift would be over. Although some people may argue that this is not stealing, the company is paying for services which it is not receiving.

When the theft of time becomes organized it can result in an entire work crew getting time off each week with pay. This can and has been carried to extremes. In one case a worker held down two jobs simultaneously, one where he actually worked and the other where he had been hired but never came to work, having a foreman clock him in and out daily. His *free* pay check was equally divided between the foreman and the worker.

A more sophisticated version of this is the hiring, by personnel employees, of a totally fictitious person. In a well-organized personnel theft ring, social security numbers are obtained, taxes are paid, employee identification cards are issued, and any number of non-existent employees, depending on the size of the company, receive pay checks every pay day.

Unbelievable? It has happened and more than once or twice. Another, less complicated version of this type of theft is to continue to issue pay checks to employees who have terminated or in some cases to former employees who have died. It is easy to see that with only ten such overpayments that average between $150.00 and $300.00 per week, the take can be sizeable.

Tighter controls which can be implemented to preclude this type of theft from succeeding include spot checks of employees who are present for work from time cards or payroll printouts. Another control might be to require all new employees to attend orientation sessions, given by security, safety and health personnel as well as selected department heads, in the first few weeks of employment, using the payroll as an attendance roster. Requiring each employee to sign for his pay check, in person, after providing identification other than a company identification card, is yet another method of better control.

Credit unions have made loans to nonexistent employees or to existing employees who not only did not receive the money but who were never aware that a loan had been made in their name. This type of activity can most easily be controlled by periodic unannounced audits which include a spot check of loan recipients by personal interview.

The possibilities are endless and much too numerous to be listed here. This only reflects part of the problem, for with every positive control initiated someone will certainly attempt to devise a method of defeating the control.

Consider the company sales store giving an employee more product than was paid for; cafeteria workers giving their friends free meals; production piece workers being paid for more production than they had produced; company vehicles being leased to third parties without the company's knowledge; padding expense accounts; issuing purchase orders for products not needed and never received. The list goes on and on. If it can be imagined, it can happen.

The potential for thievery exists everywhere and without adequate internal controls it is undoubtedly being accomplished. Conducting security/ loss prevention surveys of these controls should be a never ending cycle to insure that the controls have not been circumvented.

The impression may have been given there are no honest employees left in our work force. This is not the case. The thieves in the work force constitute a minority but still account for the loss of billions of dollars annually to American businesses. This loss hurts the honest employee most of all. It takes away income that might properly be channeled into better working conditions or fringe benefits and higher salaries.

As can be seen, there is more to loss prevention and risk management than meets the eye. It is a never ending battle of wits between protective measures and controls and those who are not satisfied to work for an honest day's work. It is also a never ending search for the most cost-effective way of conducting business.

QUESTIONS FOR DISCUSSION

1. Discuss the responsibilities that will be assigned to a risk manager.
2. What is the relationship between loss prevention and company profits?
3. What is the importance of establishing good internal controls?
4. Is it possible to maintain an effective security program without internal controls?
5. How could controls be established to minimize losses from incoming and outgoing shipments?
6. Should the theft of time be the concern of the security force?
7. Discuss some methods by which the theft of time can be detected and stopped.
8. In what manner can thefts be minimized from credit union funds?

9.   Should undercover investigators be considered if losses from theft are suspected?
10. Can security/loss prevention surveys be a one time only project and still maintain the integrity of the loss prevention program?

MAJOR CONSIDERATIONS

1.   The risk manager has a wider range of duties and responsibilities than did the position of security manager.
2.   Loss prevention techniques require the support of all company employees, from the chief executive officer to the hourly employee, if they are to be effective.
3.   Internal controls, audits, inspections and surveys are integral parts of a good risk management program which will enhance the profitability of the company.

II. PHYSICAL SECURITY: ADJUNCTS AND IMPLEMENTATION

Chapter 7

PHYSICAL SECURITY

In establishing a security system, the area to be protected must be defended by providing for absolute control of personnel and vehicular movement into and out of the area. To provide anything short of absolute control would leave gaps in the protective screen, thus rendering the entire system ineffective.

In this and following chapters, the methods, material and personnel which are required to establish and maintain the required control will be discussed.

PERIMETER SECURITY

The first line of defense in establishing the necessary control must be to provide a means by which the perimeter of the protected property can be enclosed. A simple definition of the perimeter of a protected area would be that it consists of 360 degrees of the outer edges of the property, extending outward from the hub or center of the property.

Chain Link Fence

The method most often used in providing this first line of defense is to surround the protected area with a perimeter barrier. The barrier can be constructed of any material which would deter entry or can even be a natural obstruction. The most commonly used barrier and the one recommended here as the most efficient for the money expended is chain link fencing, as shown in Figure 7-1. When used as a perimeter barrier, the chain link fence must meet the following minimum standards.

Figure 7-1.   This perimeter barrier combines block wall construction with chain link
fencing and gate. Photo by author.

The wire used should be eleven gauge or heavier and the chain link
portion of the barrier should attain a height of seven feet. The chain link
fencing material should be stretched as tightly as possible between metal fence
posts, which need to be set in concrete for stability. Posts which are used to
support the fence, at points where it changes direction (corners), require additional support such as utilizing a heavier metal which would be sunk deeper
into the ground and which would require a greater cement base.
The fence should be topped with three strands of tautly strung barbed
wire. To make entrance over this barrier more difficult, it is recommended that
this overhang be extended upward and outward at a forty-five degree angle.

When completed, the entire structure should extend to a height of eight feet. The outward angle of the overhang is designed to exclude entry. When used in facilities such as stockades or prison compounds, the overhang can be extended upward and inward.

The bottom portion of the fence should be no more than two inches above the ground; where soft soil or sand is present, it should be extended into the ground or anchored in some fashion. Special security consideration, in the form of added barriers, must be given to areas where water erosion takes place. The normal solution for this condition is pouring a concrete trough and installing a heavier metal barrier or anchoring the chain link fence into the cement of the trough. Although culverts are sometimes used, the expense is considerably greater than for the method described above. The amount of drainage that is required would be the deciding factor.

At points along the perimeter barrier where structures, trees or other elements exist outside the barrier but within sufficient proximity to be used to gain entrance over the fence, consideration must be given to increasing the height of the barrier to preclude such entry. This would include the existence of a roadway along the outer edge of the fence, where potential intruders could place a large vehicle to permit them access over the barrier.

Although the guard force might not be consulted prior to the installation of the perimeter barrier and indeed may not even have been organized at the time of construction, they will be required to patrol the barrier and they should make recommendations to management concerning any defects or inadequacies which they observe in the barrier. Items they should look for include broken or bent strands of barbed wire in the overhang, indicating entry or attempted entry; holes cut in the fencing material or existing between the fence and the ground; rusted or weakened wire in the fence and/or overhang; uprooted or sagging fence posts; and any other condition which would lessen the effectiveness of the barrier as a deterrent to entry.

The barrier that is used for perimeter protection may range from an old-fashioned moat to a prison wall or the myriad of items in-between (see Figure 7-2). The idea and the end result are the same. In the days of King Arthur and the Round Table, the castle was surrounded by a moat and access to or from the protected area was controlled by security guards who raised or lowered the draw bridge. Old fortresses were surrounded by high walls which had a parapet around the top on the inside and which were patrolled by soldiers who authorized the opening of a gate. Settlers arranged their wagons in a circle to provide a secure perimeter. Soldiers in the field have always established a secure perimeter whenever they made camp.

As can be seen, security protection has not changed much in all these years. True, we have closed circuit television and sophisticated alarm and access systems, which are in many cases computer controlled, but the principle

Figure 7.2. This eight-foot concrete block wall provides perimeter protection and privacy to residential housing area. Photo by author.

remains the same. The solid foundation of a good security program is established by controlling movement into and out of the protected area.

Gates

Just as the castle needed a draw bridge, so does a chain link fence require gates or openings. A business requires entry and exit traffic of its employees, materials, goods and, in many cases, customers. The job of the security force is not to provide total exclusion from the area, but rather to have the means of selectively excluding those persons not authorized to enter while at the same time providing a smooth flow of those persons or vehicles which are authorized to enter.

In selecting the location and determining the number of gates required for effective operation of the facility, the following factors should be considered.

1. Are pedestrian gates sufficient to handle existing and anticipated traffic?
2. Are pedestrian gates located in an area that will provide employees a direct route to their workplace?
3. Are separate gates provided for employee traffic and shipping/ receiving traffic? Is separation necessary?
4. Do gates, when closed, meet the same exclusion standard as the barrier?
5. Will temporary gates be required for future construction?
6. Will the budget provided for security services allow for sufficient guard personnel to man each open gate?
7. Will the necessary electrical power and telephone lines be installed at gate locations where guard houses will need to be installed?
8. Will one of the gate houses be used to double as the center of security operations? If so, have plans been made for the necessary wiring required for annunciator panels and closed circuit television?

There are many types of gates which can be obtained and installed to suit just about any need. Some of these will be listed below, but the supervisor should insure that his personnel are familiar with the function and operation of the gates installed on the facility where they are employed.

Pedestrian Gates or Turnstiles.   Most pedestrian gates are constructed of the same chain link material, meeting the same exclusion standards as previously described for the barrier. There are also a large variety of turnstiles on the market which may be set for one- or two-way use.

Vehicular Gates.   These gates also come in a large selection of styles and sizes. Again, they must meet the same exclusion standards as the barrier, including the provision for installation of the barbed wire overhang.

Types of vehicular gates include those that swing open in either direction or, in some cases, in a single direction; those that raise vertically; or those that roll on a track. The names given to the various types are single- and double-swing gates, vertical lift gates and single- or double-cantilever gates. Most are manually operated but can be designed for electronic operation to allow for remote control. When remotely operated, they are generally observed by a closed circuit television camera.

Any gate breaching the perimeter barrier must be controlled by a security officer when opened or securely locked when not in use. The type of lock most suitable for this function will be discussed in a later chapter dealing with locks and key control. Those gates or turnstiles which are operated in conjunction with a key card system are constantly locked and are only released when the proper magnetic key card has been inserted in the card reader.

AREA SECURITY

The second line of defense in the protection of the guarded property is the area that lies within the perimeter barrier but is outside any structures that may be on the property. This area may be unused or may be used for many diverse purposes. The normal usages that such areas may have are as follows.

Outside Storage Areas

These areas may consist of raw materials, constructed parts awaiting use in the productive process or finished products. Regardless of the type of material being stored, it has some value and in many cases it will have a high value, making it a target for theft. It is highly important that all such storage be well organized and not be permitted to become a hodge-podge of material scattered haphazardly around the area. The necessity for such organization of storage will be better understood after reading later chapters pertaining to effective lighting and security patrols.

In most situations, the material being stored will be covered for protection against the elements. If the storage area is a permanent one and has been well organized, concrete slabs will have been poured to provide a solid base for the material. Metal eyes will have been imbedded in the base, at the time of construction, to provide a tie down for the plastic or canvas covers used to protect the stored material. The use of locks or seals to insure the integrity of the covers is recommended as they will provide patrolling guards a method of determining if the material has been tampered with.

Truck Parking Areas

In most industrial and some wholesale or retail operations, trucks which were loaded near the end of the day or which were only partially loaded will be parked within the protected area overnight. Ideally, an independently fenced, impound-type area should be provided for the placement of these vehicles. If the limitations of space or budgetary considerations make it impossible to provide a separate area, then all entry doors to the trucks or trailers should be secured with a padlock or, as a minimum, a seal. Trucks or trailers having only rear entry doors can be spotted back to back to prevent the opening of the doors without moving the truck or trailer.

Employee Parking Areas

Once again, budgetary and space limitations will dictate the size and location of these areas. If possible, they should be completely outside the perimeter of the protected property. The best alternative would be to fence off the area designated for employee parking and provide pedestrian gates between the parking area and the administration/operational areas of the facility. Separate entry/exit gates should be provided for the orderly flow of traffic during shift changes. Parking lots should be paved and spaces marked to provide for maximum usage of the available space. Designating parking areas for the various production and administrative departments can be helpful in allowing employees to park in the closest proximity to the pedestrian gate nearest their place of work. This will also assist the security force in controlling the parking and in checking for valid automobile decals. Using this system, visitor parking will be easier to control since all visitor vehicles can be assigned parking spaces near the administrative offices. Again, where possible, employee and visitor parking should be maintained outside the established perimeter.

Employee Recreational Facilities

Many companies, particularly the more modern facilities, set aside areas for the recreational use of company employees and their guests. Where these facilities are located within the protected area, they must be separately fenced and provided with pedestrian gates controlled by key cards or security officers, if access to and from the area is to be allowed. By excluding access to the protected area through the use of independent entry/exit gates, unrestricted entry may be allowed into the recreational area. This permits family members and guests to meet employees in the area without the expense of providing security officers, pass systems and other restrictive measures.

In establishing a good area security program consideration must be given to what needs to be accomplished. A well-organized, clearly marked and frequently inventoried storage area adds to the efficiency and cost effectiveness of the business operation. A major concern of the security force is that patrolling or static guards have a good unobstructed view of the area and that they are able to safely patrol the area, day or night, without the possibility of incurring injury from hidden obstacles.

In order that the security force be able to adequately observe the area during the hours of darkness, an effective protective lighting system must be installed. In King Arthur's Court, torches were lit and strategically placed so that the light illuminated the moat and the courtyard outside the palace but within the wall. Due to the importance of protective lighting to the security system, an entire chapter has been devoted to its discussion.

BUILDING WALLS

The peripheral or outside walls of the building within the protected area make up the third line of physical protection. At times, building walls will form a part of the perimeter protection of the facility. In such cases any openings in the face of the building wall should be treated exactly the same as outlined for gates breaching the perimeter barrier.

In either case, whether on or within the perimeter, any opening which could be used to gain unauthorized entry must be secured when not in use. Entry through these openings must be controlled at times when entry has been authorized. As a rule of thumb, any opening which is more than ninety-six square inches in diameter, or is less than eighteen feet from the ground or is less than fourteen feet from another structure should be secured to provide adequate physical protection.

Fire regulations will require that emergency fire exit doors be either unlocked or equipped with panic-type hardware allowing the door to be opened from the inside in case of the need for emergency evacuation. To insure the integrity of these doors, they should have some type of alarm system installed. Available systems that are appropriate for this use will be discussed later in the book.

Normally, windows will be secured with the same type of chain link fencing material which is used to provide perimeter protection. Sometimes, however, metal bars or other types of screening materials are used. As a safety factor, screens may be affixed to the window in such a way as to allow them to be locked with a padlock and easily removed in the case of an emergency. In the case of windows in building walls that form part of the perimeter of the facility, consideration should be given to the installation of an alarm system.

Although the guard force will not be responsible for the installation or maintenance of such items, they are responsible for making regular checks of the protected area during the course of their patrols to insure that the devices in use are maintained in a good state of repair and that they have not been tampered with. Discovery of broken or cut screening material, missing locking devices, broken hasps, weakened frames to which screening has been attached or any condition which might compromise the security of the opening should be reported by the patrolling guard and followed up by the supervisor to insure that the necessary repairs or changes have been effected by the maintenance department.

If such protection is not presently in use at a facility, recommendations should be made, through proper reporting channels, to management outlining the importance and value of providing the necesary security screens, locking devices and alarm systems, where applicable.

BUILDING SECURITY

The fourth line of defense is devoted to special security requirements within the buildings being protected. The extent and degree of security protection which should be implemented will be determined by the activity carried on within each building in the complex. Therefore, it is necessary that each building or structure within the protected area be considered separately.

For instance, a building used solely to house raw materials such as heavy steel ingots to be used in the productive process would require a lower degree of security protection than a building housing small expensive transistors to be used in the production of communication equipment. Similarly, the portion of a building being used for research and product development would require more stringent security measures than the portion housing the sales office.

If adequate security protection is to be provided, it will be necessary to conduct a survey or study of all operations taking place in each part of the building to be protected, establish priorities based on the need for protection and the amount of money budgeted for this purpose, and plan the security protection accordingly.

Security professionals must always be aware that they are employed to protect the assets of the company for which they work. They must keep in mind that the ultimate goal of any profit-making enterprise is to maximize profits. These considerations must be part of any organizational plan implemented to achieve the security objectives. Accordingly, in planning what protection, if any, is required within a given area or building, the effect on operational efficiency must be considered. Trade offs may have to be made if a totally effective security program is to be realized, establishing the best security protection possible that is consistent with efficient operation of the business.

The added safeguards which are required in research and development departments and proprietary information storage, planning and discussion areas have already been discussed. Some examples of other areas which should be considered for special attention would be storage areas for money, precious metals, jewels, and drugs; credit unions; cafeterias; quality control departments; and any area where parts or products are stored which a pilferer would find tempting, such as those which are small, expensive and easily sold.

In each case the goal is to provide a secure area, with appropriate access controls, within the protected property. These generally take the form of vaults or security cages or excluded areas or buildings. Access to any of these areas is strictly controlled and all property which is brought in or taken out is closely documented. In addition, unscheduled daily or weekly inventories and inspections are conducted to identify any potential problems and to minimize any losses that might occur.

As was seen in Chapter 4, the size of the guard force is directly related to the effectiveness of the overall security program and in particular the physical security segment. If movement into and out of the protected area has been restricted to one, two or three points, then the number of guards required can be reduced without adversely affecting the degree of security which is achieved. On the other side of the coin, an area with no perimeter security and which lacks other physical security measures could be patrolled by fifty to one hundred security officers without achieving the degree of security protection desired.

As pointed out, the purpose of these measures is certainly not to stop movement into or out of the area since to do so would curtail or halt the business process. The aim is to control and supervise movement while allowing for the normal process of traffic flow that is essential to the operation of business be it manufacturing, transportation, retail or service oriented.

QUESTIONS FOR DISCUSSION

1. Name some types of perimeter barriers.
2. What standards should be met by chain link fencing if it is to be used as a perimeter barrier?
3. What precautions need to be taken when building walls form part of the perimeter barrier?

4. What measures need to be considered and implemented to insure proper area security?
5. What standards should be established for gates which breach the perimeter barrier?
6. Will a property protected by a perimeter barrier be secure without taking any other security measures?
7. What can be done to insure the integrity of exit doorways while still complying with fire regulations?
8. Name some areas, within the protected area, which require special security attention.
9. What are some types of vehicular gates which are used to breach the perimeter barrier?
10. Discuss the effect that meeting the physical security standards has on the structure and strength of the security guard force.

MAJOR CONSIDERATIONS

1. Providing a secure perimeter creates a sound foundation on which to build an effective security program.
2. The organization of outside storage areas makes for more efficient business operations and assists the security force in safely performing its duties.
3. Without adequate physical security measures the task of providing adequate loss prevention services is all but impossible.
4. It is important to recognize that movement needs to be controlled but not hindered or stopped.

Chapter 8

IDENTIFICATION AND CONTROL OF PERSONNEL

Establishing a solid physical security posture enables control of who is permitted to enter the protected area as well as what property is allowed to be taken out of the protected area. In order to establish the proper control of movement, management will have to answer the following questions:

1. Who is to be permitted to enter?
2. Will there be restrictions as to the time they may enter? Or, through what entrance?
3. Will there be restrictions on where they may go within the protected area?
4. Will they be broken down into classes with varying restrictions?
5. Will any employees be given keys that will permit their entry through unguarded entrances?
6. Who will be authorized to sign identification cards or make exceptions to officially promulgated policy?
7. Will visitors require escorts while in the protected area?
8. Will appropriate disciplinary action be taken on employees failing or refusing to comply with the official identification policy?
9. Will company policy on disciplinary actions be made part of the general work rules? How will this information be disseminated to employees?
10. Who will be authorized to permit the removal of company-owned property from the protected area?

Once these questions have been answered, control of movement into, within and exiting the protected area can be established. It should be noted that if a security program is to effectively assist in the prevention of loss, it must control movement. To establish control, there must be a method of identification. The more positive the method, the more foolproof the security system.

The requirement for identification has been with us as long as man has found it necessary to defend his home, work place, city or country. In the days of King Arthur, a knight identified himself by wearing the crest of the kingdom he served, both over his chest and on the face of his shield. As an added precaution he was required to raise his right hand to show it bore no arms and to lift the face plate of his armored helmet to display his facial features. This custom was later changed and became the salute which is now the customary greeting between military men.

In the United States, today, workers will often rebel at the idea of being required to identify themselves. Some workers reject the idea because they know that it will add control to their movement and reduce their ability to steal. A second group will reject the idea due to a mistaken belief that the control is simply because management does not trust the worker, per se. A third group, the followers, will reject the idea because the other two groups object. The fourth and in most cases the largest group realizes the necessity and benefits of positive identification and does not object to the idea at all.

Statistics show that about one out of ten workers in the United States steals something from his employer. The loss involved not only injures the employer but also injures the other nine workers. Before establishing an identification system, where none had been used before, it will be necessary for management to educate their employees as to the need for the system and the benefits they stand to gain as a result of the added security.

The guard force will play a very important role in convincing the worker that the system will help rather than hurt him. If the guard force has been fair and impartial in its dealings with the worker, if they have not made any special friends or favorites among the workers, then the worker will be more likely to accept the idea of being more closely regulated by the guard force. The guard's approach to everyone he contacts during the course of his daily duties must be highly professional. He must always remain aloof, be courteous at all times, never familiar, but remain firm in the enforcement of his instructions and company rules and regulations.

Management must make it clear to the employee that a good identification system is necessary to reduce the possibility of theft from outside sources by insuring that outsiders are not able to enter the protected area. It should be stressed that identification of employees aids in the quick and thorough evacuation of the facility in the event of an emergency. It should be further pointed out that by reducing the possibility of theft, sabotage or industrial espionage, the identification system helps insure the financial integrity of the company (providing job security) and also provides more positive physical protection to the employee's place of employment. Once aware of the necessity and benefits involved in an identification system, most employees will welcome it.

There are numerous methods of identifying personnel. The best method of insuring that a person is who he claims to be is by having personal knowledge of his identity. This does not mean that the guard who is to screen the personnel should get to know all of the employees by sight. This system has more than a few drawbacks. For instance, who does the screening when the regular guard has a day off, is ill or just quits? The average person can only identify about thirty or forty persons, even if he has stayed on the same shift for a long time and there was zero percent turnover among company employees. It also happens that a *known* employee who is fired for cause or becomes disgruntled is able to enter the area and do damage because the guard knew him, having seen him every work day for six months.

A system that has enjoyed some popularity, even though it is not particularly effective, is issuing an identification button to each employee. The button would normally bear the company's name and perhaps the employee's name and clock number. When this system is in use, a button that is lost or stolen is just as easily used by an imposter as it was by the person to whom it was issued. There is no way the guard assigned to check identification can tell if the person displaying the button is the same person to whom it was issued.

The foreman from each department could be required to come to the entrance to identify each person presently working under his control. This would probably work and provide good identification and control. It would, however, be necessary to pay the foreman to come in early enough to check all the workers in his department, the gates or entrances would be clogged with the various foremen and there would be a definite loss of productive work time.

An identification card, unless it meets the minimum standards which will be outlined later and is strictly and absolutely controlled, is no better than the button-type identification. It is a quirk of human nature that a firm will have their facility designed with security in mind, keep entrances to a minimum and surround the perimeter with an approved barrier but will then negate their efforts and, in effect, throw down the drain the money expended by not implementing an effective identification system.

The security supervisor and the guard force will be required to settle for the type of identification program promulgated for use by the company which employs them. However, if the program is not effective, it is their duty to advise management of the need for a change. This is not to say that the guard on gate duty should stop the chief executive officer and tell him that the identification system is ineffective. This type of communication must be channeled through the established reporting lines within the company. It is a principle of good management and business practice not to criticize a policy or procedure unless you are prepared to offer a workable solution to the problem. The paragraphs that follow will provide the elements of establishing and maintaining an effective personnel identification system.

THE CARD

A good identification system must start with the card. The card should contain the name and location of the company or facility for which it is intended to be used. It should bear a design which is intricate enough to make it difficult to reproduce. It should be printed in various colors which will be used to designate the status of the person to whom it is issued, the area he is allowed to enter or his work area within the facility. The card should be serially numbered for positive control. The identification card, where possible, should be manufactured, issued and controlled by the personnel department or the established pass and I.D. section of the security department. The card should also have been printed with an ink which will change color if erasures or ink-removing processes are attempted.

The Photograph

Each employee should have two color photographs taken during the employment process. The background color of the photograph may be used to designate the shift the employee is assigned or the time period during which he is permitted to be in the protected area. If an employee is seen in the protected area at a time not designated by the color coding, he can be challenged by supervisory or security personnel.

One photograph should be affixed to the identification card and the other maintained in the employee's personnel file or on file at the pass and I.D. section. Should the employee terminate without turning in the identification card, the second photo, plus other identifying data, can be given to security personnel to preclude the terminated employee from using the identification card to gain entry onto the protected property. In the event he should attempt to do so, he would be recognized and the identification card confiscated.

The Information

In addition to the aforementioned color photograph, a good identification card should contain, as a minimum, the following information:

1. Typewritten name and signature of person to whom issued.
2. Social security number of the person to whom issued.
3. Employee number or clock number of bearer, if social security number is not used, date of birth and a physical description.
4. Printed or typewritten name and title and signature of the person authorized to authenticate the card.
5. Expiration date of the card, if applicable.

The Cover

The cover should be of a clear plastic material which is laminated to the card with a heat process which would cause noticeable damage or destruction to the card if an attempt were made to remove the cover for the purpose of altering the card or changing the photograph. Figure 8-1 depicts an I.D. card manufacturing kit.

One major advantage of having the personnel department control the identification cards would be that at the time of termination of an employee the personnel department would be in a position to insure that the card has been recovered. Of course this could be done for the pass and I.D. section by the personnel department. If a terminating employee states that he has lost his issued identification card, he should be required to sign an affidavit that the card has actually been lost. Although not foolproof, this method is usually effective enough to stop some ex-employees from using identification cards which they have failed to turn in. If necessary, the final pay check of the employee can be held and delivered to him at the time the employee identification card is returned.

Temporary Identification Cards

Provisions must be made for allowing an employee to enter his place of work when he has lost or temporarily misplaced his permanently issued identification card. For this reason color-coded identification cards should be prepared which are plainly marked **temporary**. Should an employee attempt to enter without an identification card, he should be sent to the personnel department, where it will be verified that he is still employed and approval given for the issuance of a temporary card which must be signed for, picked up and turned in. on a daily basis. A time limit should be imposed, preventing a given employee from using a temporary card for a lengthy period of time. If the permanent card has been lost, the employee should sign an affidavit to that effect and should be issued a new identification card. The number of the old card should be provided to the security department and notices posted in all the entrances to the facility to preclude the unauthorized use of the lost card.

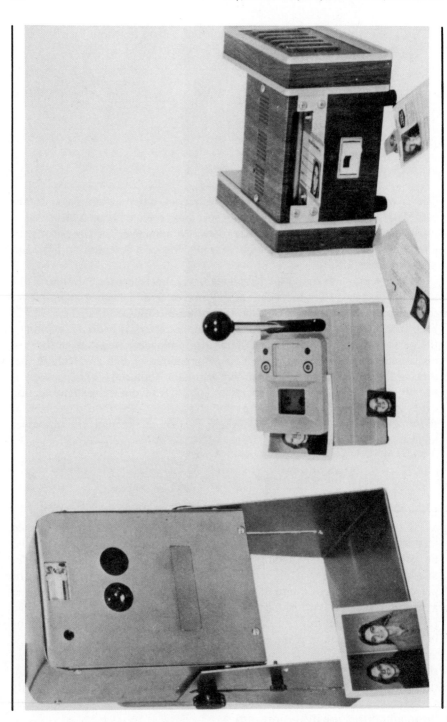

Figure 8-1. Identification card manufacturing kit designed for businesses employing 1000 persons or less. Courtesy of Identatronics.

PROPERTY PASSES

The use of lunch box and other inspections to deter the pilferer from being able to remove stolen property from the protected area has been discussed previously. There will be times when product, material or equipment will be authorized to be removed. In such cases a property removal pass should be used. These passes should be prepared, at least, in duplicate and be serially numbered for control. They should fully and accurately describe the article(s) which are to be taken, the amounts which are authorized to be removed, and should identify the person to whom the pass is being issued by name and identification number. Persons who are designated to authorize such removal should be kept to an absolute minimum and copies of their signatures should be maintained at each guard post. The guard should take a copy of the pass from the person removing the property and turn it in with his other reports. These should be returned to the person who authorized removal of the property and should be double checked to insure that nothing has been added to or changed from what was originally authorized to be taken from the protected area. As previously discussed, the package may be sealed at the time the property pass is executed. If this is done, the guard would only need to occasionally spot check packages to insure that nothing has been added to them after the seal was affixed.

VISITORS

It will also be necessary to design identification to accurately identify the various types of visitors who will be permitted to enter the protected area. It is generally desirable that visitors be broken into two major categories: those who must be escorted while in the area and those who may proceed into the area alone, except to enter those areas specifically restricted to department employees and escorted visitors. The business offices will usually receive one category of visitor and the production areas another. Most sales persons, for instance, would be interested in going to the business offices while delivery truck drivers, contractor personnel, or vendors would most likely be entering the production areas. Any visitor would be limited to that area necessary for him to perform the function for which he was admitted to the protected area. It would be preferable if visitor passes were also color coded, indicating the area where the visitor has business. In any event, all visitors should be controlled, issued an identification card, and required to sign in, return the card, and sign out when leaving the property.

Visitors to the protected area must be courteously identified and the reason for their visit established. It will be necessary for the security officer to contact the person or department to be visited to make sure that the visitor has valid business on the premises and find out if entrance for the visitor is to be authorized. Depending on the rules in force at the facility to be visited, it may be necessary for a member of the guard force to escort the visitor to the office or department he has been authorized to visit, or the person being visited may come to the guard post to personally escort his visitor.

In either event, the security officer, having established the true identity of the visitor, should issue the visitor a pass and have him record at least the following information on the visitor log sheet:

1. The visitor's name and the company he represents.
2. The reason for the visit.
3. The person or department he is visiting.
4. The time the visitor entered the facility.
5. The time the visitor departs the facility (to be filled in at the time of his departure).
6. Remarks (issuance of protective clothing or equipment which is required for entrance or any other pertinent data).

Industrial facilities, as a public relations vehicle, will often invite local student groups or other community groups to tour the protected area. It is not desirable to register each group member individually, but the person in charge of the group should be required to register and to list in the remarks section of the visitor log the exact number of persons in the group. The security officer should count the number in the group when they enter and again when they depart to insure that no unauthorized person enters with the group and that all members of the group depart the area after the tour is completed.

If members of the security force are not assigned to accompany the tour, supervisory plant personnel or adult members of the group should be designated to insure that group members do not straggle, enter unauthorized areas or come too close to moving machinery where they might be injured or inadvertently cause damage.

Every facility requires the services of vendors, utility repairmen and other contract or service personnel. If these persons are required to enter the facility on foot, they should be identified and controlled in the same manner as other visitors to the protected area. In the event these persons should be authorized to enter the facility in a vehicle, they will, of necessity, be handled differently. The manner in which they will be controlled under these circumstances will be discussed in the next chapter. In cases where the same vendor or serviceman is required to enter the protected area each day for a period of months or longer, consideration can be given to providing a more permanent, photo-bearing identification card.

CONCLUSION

In order for a card identification system to be effective, employees should be required to wear the card on a designated part of the outer clothing whenever entering or within the protected area. This rule should apply to all persons in the area whether they are management, office worker, production worker, vendor or other business visitor. Some members of management might feel that they should be recognized on sight, but they should set the example for the other employees and let them know that the rule applies to everyone and will be strictly and impartially enforced.

The day-to-day effectiveness of the identification system will depend on how thoroughly the security force checks the identification cards of employees and visitors. The photograph on each identification card must be compared with the person who is presenting it to gain entrance to the protected area. Unless the security force properly performs its duties, the entire system falls apart. No amount of planning, physical barriers, alarms or other security aids will provide the required degree of security if the identification system is ineffective. A well-trained, professional security officer will never rely on memory but will positively check each identification card.

It is good practice to periodically check on the performance of gate officers by having someone unknown to the security force attempt to enter on a forged or turned in identification card. This is particularly effective if the security force is aware that they will be so tested. The security supervisor should constantly remind the guard force that they are professionals, their responsibilities are great and their performance of duty must be flawless if the integrity of the security program is to be maintained.

Company management personnel also bear a great responsibility for the success of the identification program. In order for the program to be effective it must have total management backing. An employee who refuses to comply with the rules and regulations pertaining to identification must be dealt with in accordance with established disciplinary procedures as laid out in the work rules of the protected facility.

QUESTIONS FOR DISCUSSION

1. Why is it necessary to establish a controlled identification system?
2. Discuss the various types of personnel identification systems, their good points and their drawbacks.
3. What can the security officer do to insure that employees will willingly accept the identification system?

4. What information should be included on a good identification card?
5. How does the color coding of identification cards assist in the control and movement of personnel?
6. In addition to employees, what other classes of personnel need to be identified?
7. Since all employees are issued permanent identification cards, why are temporary identification cards necessary?
8. What method should be used to control issue and return of identification cards and what department or section should be given the responsibility for these controls?
9. What entries should be included on a visitor log sheet?
10. How important to the maintenance of the security system is the method used by security officers to check employee identification?

MAJOR CONSIDERATIONS

1. Control of movement into, within and out of the protected area is essential to any effective security/loss prevention program.
2. Movement control is impossible without a workable identification system.
3. The more positive and effective the established identification system proves to be, the greater the control of movement.

Chapter 9

IDENTIFICATION AND
CONTROL OF VEHICLES

Probably one of the first times anyone thought there was a need to establish vehicle control was after the Greek army had succeeded in breaching the perimeter security of the City of Troy. In this legend, the Trojans themselves brought about their own downfall through their perimeter defenses after their defenses had withstood the onslaught of the Greek army for ten years. Entrance had finally been gained through the use of subterfuge, causing the Trojans to relax their security.

Unless a viable vehicle identification and control program has been initiated and is enforced in any of today's industrial or other guarded facilities or complexes, thieves will certainly devise "Trojan horses" to circumvent the security of the protected area. This legend points up a very important aspect of the security officer's duties—the integrity of protection will be maintained only as long as there is no lapse in performance. Just as with the Trojans, a relaxed attitude towards security can be a fatal blow to the maintenance of a program.

Vehicles which are allowed to enter the protected area probably present the best method a pilferer or thief has of moving stolen property to a point outside, where it can be disposed of or sold. To protect the integrity of the security system, it will be essential to have an identification and control system for vehicles which is just as effective as the control established over personnel.

PRIVATE VEHICLES

Whenever and wherever possible, the entrance or parking of private vehicles within the protected area should be avoided. If private vehicles are to be allowed within the protected area, the parking areas designated for these vehicles must be segregated from operating areas of the protected property. This can be economically accomplished through the installation of chain link fencing, similar to that used for the perimeter barrier, to enclose the parking area and then breaching this fence with the necessary pedestrian-type exits.

It should be remembered, however, that any breach in perimeter security requires control. Control of vehicles and identification of occupants can be established at the entrance point of the vehicle into the protected area, or the vehicle can be identified at this point and the occupants identified when they pass through the pedestrian gates from the parking area or as they enter the building of the protected property. Whatever the situation in a particular facility, each vehicle and each person within the vehicle must be identified at some point along the line, before they enter the protected operational area of the facility.

Probably the best method of identifying and controlling privately owned vehicles is through the issuance of a decal to an employee who is authorized to bring his motor vehicle into the protected area. The decal should be displayed on the vehicle. These decals are generally constructed of weatherproof-type materials and can be luminous or nonluminous, depending upon the hours of operation at the facility. The vehicle decal should be protected from compromise in the same manner as the identification card. It should be manufactured, issued and controlled by the pass and I.D. section of the security department or by the personnel department. As with the I.D. card, the decal should have an intricate enough pattern to make duplication difficult. Once affixed to a vehicle, the decal should not be removable without destroying or defacing it in such a way as to make its reuse impossible. The decal should contain, as a minimum, the following information:

1. The company name and/or the name of the particular facility of the company where entrance is to be authorized.
2. The decal should be serially numbered for control.
3. The decal should have a vehicle control number. The serial number of the decal and the control number can be the same if desired.

Some companies prefer to use the clock number of the employee as the decal identification number. Whichever is used, the identification number should be large enough to be read from a distance.

Ideally, for ease of identification, the decal should be placed on the left front bumper of the employee's automobile. Decals, as identification cards, may be color coded. Colors may be used to designate areas authorized for parking, times authorized for entrance, or other designations decided upon by the management of the facility.

Checking the identification cards of vehicle occupants at the vehicle gate is not desirable. It is too difficult for security personnel to match the identification card, its photograph and physical description with the bearer of the card. If required, the check at this location will delay employees entering the parking areas and will often result in traffic jams on the public roads leading to the protected facility. Situations like the above result in ill will between security personnel and employees and cost the employer through the loss of productivity on the part of the employee who spends his first hour or two on shift fuming about how long it took him to clear security. Eventually it will probably result in the guard force not properly checking the identification cards of personnel entering the facility in vehicles. When this occurs, the protective screen has been breached and all the money the employer is paying for security hardware and personnel is going down the drain.

If a security supervisor is faced with a situation similar to that outlined above, he should recommend to management that changes be made in the system to permit an identification card check at a point where the employee can be processed on foot by the security guard force.

In providing the degree of security required for the facility being protected, it should be kept in mind that in most industrial or commercial operations a smooth flow of traffic is required for the profitable operation of the business. Although security procedures must be strictly enforced, they must also be flexible enough to allow the business operation to function effectively. As pointed out before, to do otherwise could result in a greater monetary loss from lost productivity than would have been lost through pilferage or theft.

Unless the physical configuration of a particular property does not permit it, visitor parking should be completely outside the protected area. Parking spaces should be provided, wherever possible, as close to the entrance to the facility as space and conditions permit. In the event visitor's vehicles must enter the facility, it will be necessary to establish some extra regulatory controls for these vehicles.

TRUCKS

At some time or another almost every area or facility will require that some vehicles be allowed to move within the protected area. Such items as telephones, electrical appliances, heating and air conditioning, lighting systems and other day-to-day necessities for the operation of a plant building or other facility will require that service personnel be permitted to bring their vehicles and equipment into the protected area.

The driver plus any assistants he may have will, of course, be required to register in the same manner as other commercial vehicles which are permitted to move within the protected area. Added control can be established by issuing a visitor card to be displayed inside the windshield whenever the vehicle is on the property. The card should be numbered for control and identification and may be color coded for the area permitted to be visited if necessary. This same card may be used for the private vehicles of visitors described above.

Whenever possible, trucks entering the facility for the purpose of picking up or delivering raw materials or finished products should be routed into and out of a separate gate from that used by other traffic (Figure 9-1). This gate should be located as close as possible to the shipping/receiving docks. All persons in such vehicles must be identified and be required to sign

Figure 9-1.   Shipping area enclosed by chain link fencing. Gate house shown is the type normally used in industrial settings. Gate is electronically controlled and could be remotely operated in conjunction with the closed circuit television camera mounted on top of gate house. Photo by author.

in on the truck log. In utilizing the truck log, certain information should be recorded by or confirmed by the security officer at the gate.

Truck Log

The truck log, in order to be effective, should contain as a minimum the following information:

1. The name of the company which owns or operates the vehicle.
2. The driver's name and the name of any assistant who would be authorized to enter the property with the driver. (Positive identification must be required of all person(s) entering the protected area through a vehicle gate.)
3. Description of the load, if delivering or picking up a partial load.
4. Destination of the vehicle within the protected area.
5. Date and time of entrance.
6. Date and time of exit.
7. An identification number or license number of the truck with a separate column for any trailers being pulled.
8. Name or initials of the guard on duty at the time of entrance/exit of the vehicle.
9. Remarks.

Truck logs will normally be printed on single sheets and maintained on a clip board for ease of handling by security personnel and truck drivers. In a normal situation the guard on duty would go out to the truck rather than having the truck driver dismount and come to the guard. It is essential that the guard inspect the truck to preclude the possibility of unauthorized personnel entering the protected area by hiding on or in the truck. If a truck is sealed when entering, the guard should record the seal number on the vehicle log, in the remarks section, and check the driver's bill of lading to verify that the seal in place has the same number as the seal placed on the truck at its point of origin.

Seals

The use of seals on trucks departing the protected area greatly assists in minimizing the potential use of these vehicles for the transportation of stolen property outside of the protected area. The shipping department foreman or supervisor should place seals on all doors of a vehicle's storage areas once it has been loaded. The number of the seal being used should be recorded on the shipping documents accompanying the load. Prior to exiting the property, the driver would be required to give the security officer the shipping documents for verification of the seal number and trailer number. Any discrepancy would require holding the truck until clarification was received. At the time of this check, the guard should record the seal number(s) on the vehicle log. Normally, it would be sufficient to conduct a visual inspection of the exterior portions of departing vehicles. In some cases it may be desirable to conduct spot checks of the interior of vehicle cabs and under the hood or in storage areas.

Movement Controls

In all cases, but particularly in cases where seals are not being used on departing vehicles, the security force will find it helpful to establish movement controls on trucks while they are within the protected area. If the shipping/receiving docks are within view of the gate, these controls are not necessary. Some controls which have been effectively utilized by guard forces and which may be adapted for almost any protected area are:

1. Installation of closed circuit television to monitor the route that trucks are to take to and from the shipping/receiving docks. Monitors for these cameras should be maintained and monitored by the guard(s) on the vehicle gate.
2. Installation of a time clock in the gate house for the purpose of stamping and issuing a time card to each entering vehicle. A second clock, located in the dock area, will permit the dock foreman to "clock the vehicle in" upon its arrival at the dock. The procedure will be used in reverse once the truck is loaded and ready to depart the dock area. This permits the time of travel to be checked to insure that no stops have been made.
3. If the route between the gate house and the docks is quite lengthy or if there are possible deviations along the route, it would be advisable to install directional signs to the docks and back to the gate to insure that vehicles do not inadvertently roam through the protected area in search of the docks or the proper gate. The extent of such signs would be determined by the scope and complexity of the area.

RAIL CARS

Rail cars entering the property will be brought in by railroad crews, spotted at the dock or along the siding and left until loaded or unloaded. Prior to movement, railroad crews should always seal each car. Security personnel should open rail gates whenever such movements are required, check to see that the cars are properly sealed, and inspect the undercarriage and exterior to prevent stolen property from being hidden somewhere on the car.

During rail car loading operations, pilferers will often drop items under the car for later pick up. Security officers, when patrolling the rail dock areas, should check beneath rail cars spotted at the dock. The same type of activity should be conducted at the truck dock to preclude property scheduled for shipment from being diverted in this manner.

ENFORCEMENT

Security officers, because of their role as enforcers of company rules and regulations, will incur the wrath of some employees they find in violation of the rules. This is particularly true in the area of vehicle movement and parking. These feelings of ill will can be minimized if, as noted in Chapter 8 on the identification and control of personnel, the security force shows no partiality or prejudice in enforcing the rules.

It is also helpful and necessary that the enforcement of company rules be predictable. If, for instance, a parking or traffic violation calls for a warning, citation, privilege suspension or the towing and impounding of the vehicle, the security officer and the employee should both know what action is called for depending upon the severity and nature of the violation.

To accomplish this, the risk manager or security supervisor must establish clear policies governing vehicle-related violations of company rules. These policies should define the action to be taken by security officers when faced with specific violations. Tow away zones should be positively and clearly marked, as should all parking restrictions, speed, direction and restricted travel areas.

Files should be established for violations and should be cross indexed by offender's name and vehicle number. Policies should include provisions for repeat violators. Disciplinary action taken as a result of vehicle violations should never be vested in the security department but should be handled in the same manner as any other violation of the company work rules or policies.

If vehicle movement control is to be effective, the rules and regulations to be enforced must be made part of the overall company policy manual. This information should be presented to each employee at the time of hire and then periodically reviewed in continuing information and training sessions for employees.

The security force should be constantly reminded that none of the procedures, policies or plans implemented to assure a good security system will do the job until and unless they perform their assigned tasks properly.

The risk manager or security supervisor will need to convince management that their concurrence and backing are necessary if these programs are to work.

QUESTIONS FOR DISCUSSION

1. What are the factors which make the identification and control of vehicles necessary to the maintenance of a valid security/loss prevention program?
2. How can control be established over the private vehicles of company employees?
3. Should private vehicle parking be segregated from operational areas? Why?
4. When using automobile decals for identification, where should they be placed on the vehicle and how are they to be controlled?
5. Will it be necessary to identify and control the vehicles of visitors entering the protected property? Why?
6. What information should be included on the truck log?
7. What measures can be taken to insure that truck traffic does not deviate from prescribed routes?
8. What precautions should security officers take to prevent pilfered property from being removed via rail cars?
9. What steps can be taken by the security force to insure acceptance of controls by company employees?
10. What role, if any, does management have in employee acceptance of vehicle control measures?

MAJOR CONSIDERATIONS

1. Well-defined and strictly enforced vehicle control measures serve to minimize the opportunity of moving stolen property out of the protected area and enhance the integrity of access control.
2. The segregation of employee and visitor parking from operational areas is vital in the control of vehicle movement.
3. Decals, seals, vehicle logs, restricted routes, closed circuit television, separate vehicle gates and proper performance by security personnel all contribute to good vehicle control.

Chapter 10

KEY CONTROL AND LOCKING DEVICES

By now, a definite picture of security should be emerging. The definition and meaning of security is very clear and very simple. Obtaining effective security, particularly for a large building, area, or complex, is not quite as simple.

As the various requirements for a viable security plan and operations are reviewed, it becomes evident that the operation requires more than a man or two to watch the premises. Rather, a multitude of requirements, all of which are interrelated, form the security system. The guard force, while a part of the system, is also the catalyst which brings all the other requirements together and molds them into an effective protective system.

An operation designed to provide security which does not incorporate all of the elements discussed in this text will not provide the degree of protection required to adequately secure the property. Any part of the overall system that is implemented will provide a degree of security protection, but unless all elements of the system are present, the protective screen will be incomplete.

Many businesses practice a false economy in implementing their security plan. Perhaps they will install a perimeter barrier and employ a less than adequate guard force in the belief that the barrier, coupled with having someone on the premises, will provide a deterrent to potential intruders. By the time they realize their mistake, the implementation of an effective system will have cost them much more than it would have to employ an effective system in the beginning.

The selection of personnel, identification of threats or hazards, loss prevention, physical protective devices, and identification and control of personnel and of vehicles have been discussed. Now a lock must be put on the protective system being designed.

LOCKING DEVICES

History

Probably the only place in the history of civilization where a lock was not needed was in the Garden of Eden. The earliest example of a lock that man has found was exhumed from the Egyptian ruins of Nineveh in ancient Assyria. This lock was found with its key and was constructed entirely of wood. It was, however, the prototype of locks still being used today, centuries later.

Biblical accounts in the Old Testament and histories of early Greek and Roman civilizations record the use of locks. It has also been reliably reported that the knights of old, when going off to war, locked chastity belts on the wives they were leaving behind. This would never work in this day of women's liberation movements, unless perhaps in reverse.

None of us are strangers to locks. We have seen and used them all of our lives. The effectiveness of locking devices in preventing unauthorized entry may need to be evaluated. Locks are said to be installed to keep honest persons honest. This saying undoubtedly came about with the realization that a dishonest person, bent on entering an area protected by a lock alone, could and would easily circumvent most locking systems.

Types of Devices

It is necessary to insure that the guard force is made familiar with the various locking devices and systems which are in use in the facility. They should be able to evaluate the deterrent value of each device and make recommendations for the replacement of devices which are known to be ineffective in preventing unauthorized entry.

The following is a listing of the various types of locking devices which are available for use, including an estimate of the effectiveness of each type of device in preventing the entry of a potential intruder.

The Warded Lock. This lock offers no delay whatsoever to the intruder and probably will not be encountered as part of a protective system.

The Disc Tumbler Lock. This lock should never be installed in any location which requires any degree of security protection. It is estimated that the maximum time required to gain entrance when this device is in use is five minutes. This device will normally be found installed in automobile doors, trunks and glove boxes. It also may be used to secure desk drawers and non critical filing cabinets.

The Pin Tumbler Lock. The effectiveness of the pin tumbler lock can vary from poor to average, depending upon the number of pins which have to be contacted by the key and the materials and tolerances used or allowed in the manufacture of the individual locks. These locks are found in general use in both industrial and residential structures.

The Circular or Barrel Keyway Lock. This device is extremely difficult to pick; therefore, it provides an above-average delay time. This lock has many applications and is recommended for inclusion in the protective system.

The Lever Lock. The effectiveness of the lever lock is determined primarily by the quality of the materials used and the precision of manufacture. Heavy duty or high grade lever locks may be used to secure safety deposit boxes, while the low grade product of this type may be used on chests, cabinets and some desk drawers.

The Maximum Security Lock. This lock is produced only by Sargent & Company. Maximum Security is a trade name which was given to this locking device following its development in 1965. This lock incorporates the pin tumbler principle but utilizes three rows of pins which were designed to interlock in the keyway. This lock has proven to be extremely pick resistant and is also recommended as being an excellent choice for inclusion in a protective system.

The Combination Lock. This device offers a good delay time since all but the most expert intruder will be unable to easily find the correct series of numbers required to open the lock. To be most effective, the opening of the lock should require the use of a minimum of three numbers, and the lock should have the capability of having its combination of numbers changed without requiring a great deal of time or the services of a locksmith.

The Cypher Lock. This lock provides excellent security in that it is extremely difficult to defeat. As with the combination lock, there is no key to issue or to be compromised. It is operated by pushing a preset combination of buttons which are either numbered or lettered. The sequence and combinations that must be used can be changed in minutes without the necessity of disassembly. This type of device is recommended for areas requiring a high degree of security where turnover of authorized personnel may require frequent changes in the settings which permit entry.

It should be remembered that gaining unauthorized entry is not limited to defeating the locking mechanism being used. The door on which the lock is to be used must be strongly constructed. Hollow doors, doors with thin laminated panels or doors having glass panels should not be used.

Hardware

The hardware being used in conjunction with the locking mechanism must also be carefully selected. The use of a dead bolt locking device is recommended for the greatest protection against unauthorized entry. The bolt should be of sufficient length to provide for a solid seat within the door jamb. This should prevent entry through the door by use of spreading devices which separate the door from the door jamb to the point where the average bolt would be rendered ineffective. The door jamb itself should be checked to insure that it is well constructed and will withstand attack designed to defeat the lock and gain entry.

Padlocks

Padlocks will be in use on gates which breach the perimeter barrier, to secure storage shacks and outside storage, on overhead doors such as those used on shipping/receiving docks, and for many other purposes throughout the protected area. Padlocks can be obtained using any of the locking mechanisms described, with the possible exception of the cypher mechanism. The delay time would depend on the type of locking mechanism selected as well as the following factors:

1. The metal and hardening process used in the construction of the shackle — a soft metal would be easily defeated by the use of bolt cutters attacking the shackle, whereas a hardened steel shackle would provide a much greater delay time.
2. The metal and construction of the body of the lock — a poorly constructed padlock can be easily defeated by an attack on the body of the lock.
3. The material which is used in the construction of the hasp and the manner in which the hasp has been installed—a weak hasp will often be defeated by merely breaking it with a jimmy or a hammer. The hasp which is not bolted to the place where it is anchored and welded to prevent the removal of the bolts will often be the victim of an attack designed to remove it from its position.
4. A padlock should be designed so that both ends of the shackle are secured within the body of the lock when the lock is in the closed or locked position. A padlock which only secures one end of the shackle is easily defeated.

You can bar the barn door, preferably before the horse gets out; you can lock something in or someone out. The locking system used in providing protection should be capable of working both ways. It should both keep unauthorized persons out and provide sufficient security to insure that property or personnel will be unable to exit the facility through unauthorized openings.

KEY LOCKING DEVICES

By far, the vast majority of locks employed to secure closed gates, doors or windows will be key operated. However, these devices are used merely as delaying tactics. Unless they are used in conjunction with other measures established in a security program, they are ineffective. It must also be recognized that even when locking devices are used with other protective measures they are only able to provide a degree of protection and will not deter the most persistent intruder.

Consider the padlock and chain which is used to secure a little–used gate at the rear of the facility. There is no alarm system installed and unless the fence line and gates are patrolled and regularly checked by the guard force, a thief, saboteur or industrial spy could easily defeat the lock and gain entry without detection. For that matter, without other precautions and control measures the fence itself is no real deterrent to the persistent intruder.

However, the greatest weakness in any key locking system is the key. Using the same example as before, the intruder has a key which opens the padlock on the back gate and he locks it behind him. Now, a patrolling guard would see nothing out of the ordinary when checking the back gate. Depending upon the extent of the area, the strength of the guard force and other factors such as strategically placed alarm systems or closed circuit television cameras, this intruder could conceivably come and go through the gate and remain undetected for whatever time he required to steal or destroy.

This could be accomplished even though the guard force was employed and was following instructions to the letter. Unless the guard force was aware that the key to the lock had been compromised, they would be busy with other duties and, in the absence of suspicion, would not waste time looking for an intruder, having found that their perimeter was secure.

KEY CONTROL

Before an effective key control system can be established, every key to every lock that is being used in the protection of the facility and property must be accounted for. Of course, if each key in the system can be identified and accounted for, key control has already been established. If no formal control has been maintained in the past, it is more likely that it will not be possible to account for each key. Chances are good that it will not even be possible to account for the most critical keys or to be certain that they have not been copied or compromised. If this is the case, there is but one alternative—to rekey the entire facility.

Once an effective locking system has been installed, positive control of all keys must be gained and maintained. This can be accomplished only if an effective key record is kept. When not issued or used, keys must be adequately secured. A good, effective key control system is simple to initiate, particularly if it is established in conjunction with the installation of new locking devices. One of the methods which can be used to gain and maintain effective key control follows:

1. Key cabinet—a well-constructed cabinet will have to be procured. The cabinet will have to be of sufficient size to hold the original key to every lock in the system. It should also be capable of holding any additional keys which are in use in the facility but which are not a part of the security locking system. The cabinet should be installed in such a manner so as to be difficult, if not impossible, to remove from the property. It should be secured at all times when the person designated to control the keys is not actually issuing or replacing a key. The key to the key cabinet must receive special handling, and when not in use it should be maintained in a locked compartment inside a combination-type safe.

2. Key record—some administrative means must be set up to record key code numbers and indicate to whom keys to specific locks have been issued. This record may take the form of a ledger book or a card file.

3. Key blanks—blanks which are to be used to cut keys for issue to authorized personnel must be distinctively marked for identification to insure that no employee has cut his own key. Blanks will be kept within a combination-type safe and issued only to the person authorized to cut keys and then only in the amount that has been authorized by the person responsible for key control. Such authorization should always be in writing, and records should be maintained on each issue which will be matched with the returned key. Keys which are damaged in the cutting process must be returned for accountability.

4. Inventories—periodic inventories will have to be made of all key blanks, original keys and all duplicate keys in the hands of the employees to whom they have been issued. This cannot be permitted to take the form of a phone call to an employee, supervisor or executive asking if they still have their key. It must be a personal inspection of each key made by the person who has been assigned responsibility for key control.

5. Audits—in addition to the periodic inventory, an unannounced audit should be made of all key control records and procedures by a member of management. During the course of these audits a joint inventory of all keys should be conducted.

6. Daily report—a daily report should be made to the person responsible for key control from the personnel department, indi-

cating all persons who have been terminated or who will be leaving the employ of the company in the near future. A check should be made, upon receipt of this report, to determine if the person named has been issued a key to any lock in the system. In the event a key had been issued, steps should be initiated to insure that the key is recovered.

Security force personnel will normally be issued master keys, when such a system is in effect, or they will be issued a ring of keys permitting them to enter any part of the guarded facility. Keys issued to the security force should never be permitted to leave the facility. They should be passed from shift to shift and must be receipted for each time they change hands. The supervisor must insure that all security personnel understand the importance of not permitting keys to be compromised.

A lost master key compromises the entire system and results in the breakdown of the security screen. Such compromise will necessitate the re-keying of the entire complex, sometimes at a cost of thousands of dollars.

If re-keying becomes necessary, it can most economically be accomplished by installing new locking devices in the most critical points of the locking system and moving the locks removed from these points to less sensitive areas. Of course, it will be necessary to eventually replace all the locks in the system, but by using the manner just described the cost can be spread out over several budgeting periods.

QUESTIONS FOR DISCUSSION

1. Can locks alone provide adequate security protection?
2. What other elements have you studied which must be combined with locking devices to establish a viable security system?
3. Name some ways a locking system can be defeated.
4. What is the best type of lock to use for securing of doors?
5. What type of locking devices would normally be found on perimeter gates?
6. What is the first step in establishing a good key control system?
7. What is an economical method of rekeying a facility should it be necessary?
8. Once all keys are accounted for, what is the next step in establishing an effective key control system?
9. Do security officers have any responsibility for insuring the integrity of the key control system?
10. What are the ramifications of a lost master key?

MAJOR CONSIDERATIONS

1. Locks do not prevent entry; they only delay entry.
2. The weakest link in any key locking system is the key itself.
3. The best locking system is only one part of the total security system, yet an ineffective locking system breaches the entire system.

Chapter 11

PROTECTIVE LIGHTING

The need for adequate lighting to permit the security force to perform their duties effectively as well as safely was briefly discussed. The installed perimeter barrier may be constructed of the highest quality chain link fencing which meets all of the standards that are required for a secure perimeter fence. The housekeeping in outside areas may be outstanding, and storage areas may be neatly organized to give the patrolling security officer an unobstructed view of company property between the buildings and the perimeter barrier. Then, darkness falls and the two 60-watt light bulbs which have been installed on either end of the 3,000-foot building are turned on. They will make about as much of a dent in the darkness as a marshmallow would make striking an anvil.

The cover of darkness has long been considered prime time for criminals and sneak thieves. Throughout the history of mankind, darkness has been used to cover covert activity. Military tacticians and criminals have used the darkness to great advantage and still do. To have adequate defenses against intruders successfully attacking the protected property, provision must be made for the elimination of darkness to take away its mantle from any potential intruder.

No one will expect the supervisor or the members of his force to be illumination engineers, to know how many light standards are needed to provide adequate lighting to the perimeter barrier, or to provide for good night-time area security. Members of the security force, particularly the supervisor, should be familiar with the basic concepts of security lighting, what is available to do the job and where to look for the technical expertise to bring the lighting up to the level required.

A little common sense, coupled with the use of one's senses, can determine if the lighting is adequate or if a recommendation for additional lighting is needed. Select a dark, cloudy or moonless night and walk the normal patrol route used by security officers in conducting area and perimeter security patrols. It would be helpful to have a diagram of the area on which to note deficiencies or comments to be brought up in the next security meeting.

Look for and record any areas where the shadows are so heavy as to preclude the security officer from being able to detect a potential intruder. Check to see if any kind of vegetation, shrubbery or heavy growth exists that an intruder could use for cover. This check should include an area of at least twenty feet on the outside as well as the inside of the perimeter barrier. If there are areas of the perimeter barrier, or the outside clear zone, that are too heavily shadowed or insufficiently illuminated to permit proper observation, then the lighting system in use is not adequate.

The next step in checking the sufficiency of security lighting is to patrol the perimeter barrier, using the outside clear zone if possible, while another member of the security force follows in parallel but using the normal patrol route taken by patrolling officers. If the lighting system is effective, the patrolling officer should be able to observe your movements at all times while the lights should glare in your eyes, preventing you from observing the officer on his patrol route.

It may be that a minor adjustment in angle of existing lights will improve the lighting system or perhaps a higher intensity lamp will eliminate an existing problem. If, however, the lighting system is inadequate to the point where any changes the company maintenance personnel could make would still fall short of the necessary illumination, then expert assistance is required.

Many power companies around the country will provide, free of charge, a survey of lighting needs by an illumination engineer. Also, as a service to industrial customers, the power company will install the necessary standards and lamps on a low-cost, lease basis. If this service is not available, it would still be beneficial to obtain the services of an illumination engineer and to have adequate lighting standards and lamps installed.

CONTROL

There are three basic ways of insuring that the protective lighting system is turned on and off at the proper time.

1. Manual operation of the on/off switch at designated times—manual operation responsibility should be vested in either maintenance, security or management personnel.
2. Timing device operation in which a timing device is set to turn the lights on/off at designated times—in using this control method it should be remembered that timing devices must be changed to reflect the changing sunrise/sunset times throughout the year. This not only results in more effective security lighting but also conserves energy during the longer days of the year.

3. Photoelectric cell operation by which the photoelectric cell senses the intensity of light so that as darkness increases the control activates the light switches—conversely, as the day dawns, the photoelectric cell senses the increase in light intensity and the control turns the light switches to the off position. This is the best method of control, but a manual override should be installed in conjunction with this system in the event of a system failure.

ALTERNATE POWER SOURCES

A good protective lighting system must be able to provide the necessary illumination during all periods of darkness and under any conditions. To make this possible, a back-up power source is required to provide this necessary power should the facility suffer a commercial power outage. This back-up can consist of storage batteries or portable, gasoline or diesel powered generators. It should be pointed out that these generating systems supply direct current and are not adaptable to providing service to series lighting systems or to multiple lighting systems having controlling devices which are designed to operate only on alternating current.

Any alternate power source should be designed so as to automatically take over and provide power to the lighting system in the event of a utility power source failure.

VAPOR LAMPS

Although both mercury vapor lamps and sodium vapor lamps provide good illumination and the sodium vapor is very good in foggy areas, they are not recommended as suitable for installation in a protective lighting system. After a power failure, due to a warm up requirement before they will light, either of these lamps would require a minimum of five minutes to return to use. This would give potential intruders sufficient time to breach the perimeter, particularly if the intruders had noticed the lamps in use and had caused the power failure to give themselves the covering darkness.

LIGHTING DEFINITIONS AND TERMS

1. Candle power is the degree of light obtained by one or more candles. Thus, 5 candle power would be the amount of light given off by 5 candles.
2. Foot candle is that amount of light that illuminates an area one foot away from the light source of the candle.

3. Horizontal illumination simply refers to the amount of illumination that is received at ground level or horizontally from the light source.
4. Vertical illumination refers to the amount of light that is received on a vertical surface from the light source.
5. Continuous lighting consists of a series of light standards and lamps which provide a continuous illumination of a given area.
6. Stand-by lighting is that provided by batteries or generators to be used in the event of a power outage.
7. Portable lighting is that which, as the name implies, can be moved from location to location. These can be vehicle–mounted or hand–held searchlights, spotlights or even flashlights.
8. Emergency lighting could be a combination of any other light source and can be any type of lighting which is used in the event of an emergency requiring supplemental lighting or replacement lighting.

GUARD FORCE RESPONSIBILITY

The guard force should not be responsible for the maintenance of the protective lighting system but should routinely record on their reports the location of any light in the system, or within the protected area, which is not operating properly. This report can then be directed to the maintenance department for repair or replacement, thus insuring the integrity of the security system as well as the safety of the employees.

OTHER LIGHTING CONSIDERATIONS

The security force must also be concerned with the lighting over entrances and exists, both from the buildings and through the perimeter barrier. They need to insure that additional lighting is provided for security cages and sensitive areas such as loading docks and warehouse areas.

In addition to its importance to the security function, proper lighting is also required to insure employee safety, mark fire exits and illuminate emergency evacuation lanes. In the event of a power outage, these systems must also be included in the emergency lighting system.

Members of the security force who are assigned to patrol duties should be equipped with portable hand–held lights which emit sufficient illumination to probe dark areas along their patrol routes. This can eliminate the possibility of the patrolling officer being surprised or attacked by an intruder concealed in dark or shadowy areas.

Portable hand-held lights are available on the market which can also be used as a defensive wand or night stick. This type of light will provide the patrolling officer the additional protection of having an effective defensive weapon should he be attacked by an intruder.

QUESTIONS FOR DISCUSSION

1. How does darkness affect the efficiency of the security system?
2. Is it possible to check the illumination system without calling in an illumination engineer?
3. When is the best time to conduct a check of the security lighting system?
4. To what extent should a clear zone extend inside and outside of the perimeter barrier?
5. What organization should be contacted for professional advice on improving the protective lighting system?
6. What are the three basic methods of controlling a protective lighting system?
7. Which method provides the most reliable and economical system of lighting control?
8. Should it be necessary to have an alternate power source available?
9. Do vapor lamps provide a good light source for a protective lighting system? Why?
10. What responsibility does the guard force have for insuring the integrity of the protective lighting system?

MAJOR CONSIDERATIONS

1. Professional advice pertaining to a protective lighting system can normally be obtained by contacting the local utility company.
2. A back-up power source is essential to the integrity of the protective lighting system.
3. Security force reports pertaining to non-functioning lights within the protected property are essential if the protective lighting system is to remain effective.

Chapter 12

ALARM SYSTEMS

Man has used signals to warn of impending danger since the first cave men banded together for their mutual protection. The merry men of Sherwood Forest, American Indians and many frontiersmen both communicated and sounded an alarm by using animal or bird sounds. Paul Revere received the alarm by lanterns placed in a bell tower. One if by land, two if by sea.

By the time World War I came along, man had become more sophisticated and was hanging tin cans on the concertina or barbed wire to warn him if anyone attempted to cut the wire or crawl under the wire. World War II brought the wail of sirens alerting and warning of an impending air raid.

After World War II ended, the crime rate in this country started to outgrow the effectiveness of law enforcement agencies. Businesses found that they had to see to the security of their own property. Many businesses simply kept the guard force and security structure they had been required to maintain while in the production of war materials.

Anti-intrusion and fire detection alarms had been developed and refined as a result of the tightened security requirements brought on by the war. As the market for such systems grew following the war, these devices were even further refined. Today, there is an alarm to protect anything from a music box to a large industrial complex or an atomic power plant.

TYPES OF ANTI-INTRUSION ALARMS

Pressure Alarms

Pressure alarms, usually produced in the form of a mat, signal the presence of an intruder when a preset amount of weight is placed on the mat. Pressure-type alarms can be effectively used inside and parallel to the perimeter barrier to alert the security force that the barrier has been breached. They may be used to detect unauthorized entry into a building by placing them below windows and inside doorways. Pressure-type alarms are also useful in keeping anyone from getting within arms reach of a valuable painting or other art object.

Magnetic Contact Alarms

Magnetic contact alarms activate the alarm when the magnetic field set up between two contact points is broken, as the contacts separate when the door or window is opened. This type of alarm is both popular and effective subject to the restrictions on alarms, in general, which will be discussed later in this chapter. One contact is mounted on the door or window while the other is aligned with it but placed on the door frame or window sill or imbedded in the floor when installed on overhead doors.

Motion Detection Alarms

Motion detection alarms work on the principle that any motion in an area covered by the system will upset that balance that has been established and cause the alarm to be given. There are two basic motion detection systems available — ultra-sonic and radio frequency.

The ultra-sonic system operates by setting up a pattern of sound waves between two speakers which blanket the entire area to be protected. Any movement within the area causes the sound patterns to be distorted, which results in the alarm being given.

The radio frequency system is similar, but sets up a pattern by transmitting radio signals between two antennas. As with the ultra-sonic system, any movement in the area being protected causes an imbalance and results in the alarm being given.

One art gallery, in the mid-south, installed an ultra-sonic system in the area where their most valuable paintings were hung. The system kept giving off false alarms. Adjustments to the sensitivity of the system were made but the false alarms continued, except at settings which rendered the system ineffective. Finally, after checking out the entire system, the offending intruder was identified. Each time the air conditioning system was activated, the leaves of a plant within the area would move with the breeze and set off the alarm.

Audio Detection Alarms

Audio detection alarms consist of a series of microphones which are installed in an unoccupied area. Any sound within the area where the system is installed will result in the alarm being given. These devices are also subject to false alarms due to their sensitivity and their ability to pick up sounds which originate outside the protected area.

Photo-Electric Alarms

Photo-electric alarm systems operate by means of a light beam which is sent from a transmitting cell to a receiving cell. Any object or person passing through the beam interrupts the signal and sets off the alarm. These systems were widely used to open doors to business places prior to the take over of pressure mats as the primary means of automatic door opening.

Metal Foil

Metal foil is the most frequently used alarm system for the protection of glass doors and windows. Any break in the foil results in the alarm being given. This system is subject to many false alarms and problems in setting the alarm caused by foil broken by age or accidentally scratched while the alarm is not activated.

Vibration Alarms

Vibration alarm systems may be used for door and window protection as well as for object protection. These alarms are preset for a degree of vibration which, if exceeded will cause the alarm to be given. These alarms are not very effective in areas where heavy truck traffic is present, where earth tremors are commonplace or in areas where jet aircraft frequently exceed the sound barrier.

Capacitance Alarms

Capacitance alarms will normally be used in the protection of an object such as a particular file cabinet or safe. The protected object must be metal in order for this system to be useful. It operates by setting up an electro-magnetic field around the object, usually at a distance of 10 to 12 inches, which if entered disrupts the field and causes the alarm to be given.

METHODS OF MONITORING ALARM SYSTEMS

In describing the operation of the various alarm systems, in each case it was indicated that an alarm is given. It does not matter if the alarm is an anti-intrusion device or a fire protection alarm device (which will be discussed later in this chapter). There are four different methods by which the alarm can be given and received.

Local Alarms

Local alarms will have a bell, siren or other loud noise built into the alarm itself, which will sound only in the area of the alarm location. This type of alarm is good, providing there are employees or security personnel within ear shot of the alarm.

However, it is often the case that the alarm will be sounded at times when there is no one in the area to hear its warning.

These alarms are normally battery operated and in many cases the batteries are allowed to run down, which totally eliminates any usefulness the alarm might have. In other cases the alarm is sounded, but before anyone comes within earshot the batteries die and the alarm is silenced.

Proprietary Station Alarm Monitors

Proprietary station alarm monitors are installed on the protected property. These monitors are usually located within the main guard force office or station. When an alarm has been activated, the signal given off is transmitted to an annunciator panel which is monitored by security personnel. Most annunciator panels have an audio alarm to alert monitoring personnel and also light up a coded square on the panel which indicates the location, by area, of the alarm which has been violated. A member of the security force can be dispatched to check out the alarm, and since the alarm is silent at the point of entry, the intruder may not be aware that he has set off an alarm.

Central Station Alarm Monitors

Central station alarm monitors are installed at the office of the servicing company. When an alarm has been activated, the signal is sent over telephone lines to the central location and picked up by annunciator. These panels will usually only reflect that an alarm has been activated at a coded location tied into their system. The name and address of the protected property must then be determined and response to the alarm is started.

Most central station alarm companies employ security personnel who are assigned to various areas of the locality and who are dispatched to answer alarms. After dispatching the alarm runner, if the protected facility has proprietary guards, the alarm company will telephonically inform them of the receipt of the alarm. Local law enforcement agencies will also be notified of the alarm. Many local police departments will not respond to the alarm until its validity has been verified by the company's security personnel. In any case, valuable response time is lost, which in many cases permits the intruder time to take what he came for and make good his escape.

If the alarm received at the central station were from a fire alarm, then central station personnel would notify the fire department in addition to dispatching their own response and notifying any proprietary force on the property. Central station alarm companies provide a valuable service to home owners and small businesses both, for protection of their property against the threat of fire as well as burglary or theft. Anti-intrusion alarms can be silent at the protected property or can alert the central station while sounding a local alarm to attract attention to the intrusion. The audible alarm also serves to psyche out the intruder and hopefully to induce him to leave the property immediately.

Auxiliary or Remote Alarm Monitors

Auxiliary or remote alarm monitoring is done by having the signal relayed to an annunciator panel at the local police or fire department headquarters. As with central station alarms, the signal must be transmitted by telephone lines. This method is not available to residential or commercial users in most parts of this country since the governmental departments have neither the time nor the manpower to monitor the alarms.

Choice of Monitoring System

There is no doubt that in an industrial, medical or business complex that employs a proprietary or contract security force, the best monitoring system is the proprietary or in-house system. The quickest, most dependable alarm system manufactured is only as good as the capability to respond to the threat of fire or intrusion. When an alarm is received on a proprietary annunciator panel by a member of the security force on the property, the response can be immediate.

Annunciator panels, at either the central station or on site, have indicator lights that allow the monitor to determine if the alarm system is on line, is on line but one or more of the circuits in the system are not engaged, or the system is off. When activating an alarm system which indicates that an area or specific location does not have the alarm circuits closed, the monitor must dispatch a patrolling officer to check the area or location, determine why the circuits are not on line, and bring them on line or call for a technician to repair the problem. Members of the guard force will not normally possess the technical know-how to make alarm repairs or to take any action other than closing an open door or window to allow the circuits to close.

EMERGENCY POWER SUPPLY

As discussed with the protective lighting system, the alarm system also requires uninterrupted electrical power to remain effective. Most alarm systems can be powered, for four to twelve hours, by electrical current provided by either dry cell or wet cell storage batteries. In the event of a power failure, the alarm system should be automatically switched to the emergency power source.

With the exception of battery-powered local alarms, all other alarm systems discussed require electrical power. Therefore, it would seem a simple solution for the potential intruder to identify the wiring that services the alarm system and cut or break that wiring to prevent the alarm signal from being transmitted. To avoid the alarm system from being rendered ineffective in this manner, a line supervisor should be installed. This is accomplished by maintaining a constant flow of low voltage current passing through the line. If an attempt is made to cut the wiring or to by-pass the signal, the voltage level on the line is changed and an alarm is automatically triggered, indicating a tampering or problem with the electrical system supporting the alarms.

FIRE PROTECTION ALARMS

A good fire protection alarm system must contain three basic elements to insure immediate detection and response in the event of fire. This is necessary to minimize property damage and, more importantly, to save lives.

Manual Fire Alarms

Manual fire alarm stations, similar to those which are in use on public thoroughfares, must be well marked and easily accessible to all workers within the area they are designed to cover. As with the other types of alarms which will be discussed, these alarms may be coded to a designated fire area or zone.

Fire Detecting Devices

Fire detecting devices or alarms come in a variety of types, some of which are suitable for general use and others which are limited to certain areas of usefulness. Some of the types which may be installed in a plant or complex are as follow.

Smoke detecting alarms. These alarms generally operate on the same principle as the photo-electric cell alarm described earlier in this chapter. In the case of the smoke detecting version, it is much more compact and the distance between the light source and receiver is minimal. Smoke coming between the light source and the receiver disrupts the beam and causes the alarm to be activated. Smoke detectors are most effective when used in air conditioning or air vent ducts. This type of detector has become very popular as a residential fire alarm.

Ionization detector. This alarm operates by a small current being passed through the air between two plates. Materials such as hydrocarbons and other products being given off by a fire in the incipient stage disturb the current and cause the alarm to be activated.

The ionization detector will most likely be installed in a computer room, white or sterile room or in areas where the possibility of spontaneous combustion exists. The early warning capabilities of this type of detector make it an excellent choice for inclusion in any fire protection system.

Heat Sensing Detectors. There are several types of heat sensing detectors which operate either by a rapid change in temperature or by a rise in temperature above a given level. In either case, when the change in temperature occurs, the alarm is activated.

Supervisory Alarms

Supervisory alarms are operated in conjunction with or in support of the installed sprinkler system.

1. The water flow alarm is installed in the main riser and consists of a paddle tied to an electrical switch. When the sprinkler system is activated, usually by a ruptured head in the wet system, the paddle moves with the water flow, and its movement switches on the electrical contact which in turn signals the alarm to the annunciator panel.

2. The air pressure supervisor activates an alarm when the air pressure in a dry system or in a storage tank drops below a predetermined level.
3. The water level supervisor signals the annunciator panel whenever the water level in a storage tank falls below the required level.
4. The temperature supervisor is activated when the water in a supply tank or in the sprinkler pipes is near freezing.
5. The post indicator valve supervisor signals a warning whenever the main water valve to a riser has been shut off.
6. The gate valve supervisor sends a warning to the annunciator panel whenever the secondary water supply valve has been closed.

As stated before, and will be stated again, all the security hardware, alarms, fences or other devices will be of no avail unless the security force is well trained and is properly motivated to carry out their assigned tasks to the best of their ability.

QUESTIONS FOR DISCUSSION

1. Name the types of alarm systems in general use to detect intrusion.
2. Are local alarms effective in all cases?
3. Is it possible to shut off an alarm system by cutting the electrical power?
4. What is the best method of monitoring alarms for a facility that employs an on-site security force?
5. What is the major problem in relying solely on central station monitoring?
6. What are the three basic elements that make up an effective fire protection alarm system?
7. What is the function of the water flow alarm?
8. What purpose is served by the air pressure supervisor?
9. What function does the post indicator valve supervisor perform?
10. What is the security officer's role in insuring the integrity of the installed alarm systems?

MAJOR CONSIDERATIONS

1. Without adequate and timely response, alarms are not effective, and the money expended for their installation has been wasted.
2. Where a security force is employed on the premises, the best response can be obtained by utilization of a proprietary alarm system.
3. A good fire protection alarm system requires local alarm boxes, smoke and fire detection devices, and supervisory alarm systems installed to insure the integrity of the sprinkler system.

Chapter 13

ENTRY CONTROL SYSTEMS

It has already been established that to have an effective security program, entry into the protected area must be controlled. It is also necessary to control movement within the protected area in addition to controlling access to those areas which have been designated restricted.

The use of a good identification system and how, when such a system has been properly implemented, it can establish the necessary entry and movement control has already been discussed. The basic identification system can also be extended to positively control access to restricted areas. This can be accomplished by instituting the exchange system or the multiple system of personnel identification.

EXCHANGE OR MULTIPLE SYSTEM

The exchange system requires that a duplicate identification card be made that will be used to permit entry into the restricted area. The duplicate card will be color coded for the specific area where it is to be used, but otherwise it will be identical to the card which has been issued to the employee to permit him to enter the protected area.

When an employee is authorized to enter a particular restricted area, for instance, the research and development department, he presents his issued identification card to the guard on duty at the entrance. The duplicate identification cards are maintained by the security officer at that location. The security officer takes the issued identification card and exchanges it with the duplicate card which the employee must wear while in the restricted area. When the employee's business is completed within the restricted area, he returns the duplicate identification card to the security officer in exchange for his issued card.

The multiple system uses the exchange system when more than one restricted area is contained within the protected area. In this case the exchange system is used in each area that is designated as restricted.

Some examples of where the exchange or multiple systems might be used would be research and development, computer room, tape library, finance centers, proprietary marketing information file rooms, drug storage areas and executive offices.

Probably the biggest drawback of the exchange or multiple systems to control entry into these areas is that they require the employment of security personnel to administer the system. For most businesses the cost of using the system is prohibitive unless the degree of security requires visual inspection of all persons authorized to enter the restricted area.

ALTERNATIVE SYSTEMS

Fortunately, there are some alternatives to using these systems. A separate sub-mastering system might be installed and keys issued only to those persons authorized to enter. This alternative is not recommended. Its weakness lies in the possibility of keys being lost or duplicated, thus compromising the system and necessitating re-keying the entry point and re-issuing keys to each authorized employee.

Using an electrical strike could provide the necessary degree of security but would still require an identification procedure and would also need the same amount of manpower that would be needed for the exchange system.

Use of an electric strike, remotely operated by a security officer in conjunction with closed circuit television, is another alternative. With this system the security officer will be able to check the person presenting the card with the photograph contained on the card. Although this method does not require employing one security officer for each restricted area, it does not provide adequate identification of the person desiring entry.

Since most closed circuit television systems in use in industrial locations are black and white, the identification of a color coded identification card is not possible. Although an alternate method of identifying persons authorized to enter the restricted area can be devised, this method is not the best available for the money expended.

A greater degree of control as well as convenience can be obtained by installing the cipher or combination-type door lock. In implementing this system, only those persons authorized to enter would be given the combination necessary to activate the locking device. The only drawback to this method is that the combination must be changed regularly, particularly whenever a person having knowledge of the combination is no longer authorized to enter the particular restricted area.

MAGNETIC KEY CARD SYSTEM

The magnetic key card system is, in the author's opinion, the best method of movement control into and out of restricted areas. Such a system properly installed and fully utilized is also the most economical method a company can employ. This system utilizes a plastic card containing thousands of magnetic bits or particles which can be arranged to match the required pattern set up in the card reader. When the match is made, the system is activated to perform whatever function it is designed to do.

There are similar, although much simpler, versions in use to permit entrance into parking lots (Figure 13-1) or parking garages. The same principle is used for the cards required to operate the instant mechanical banking tellers which are being installed and used throughout the country. In this case the card opens the unit for business, but unless the person presenting the card keys in a coded number known only to him and the bank, no business can be transacted.

The simplest method of using a magnetic key card is to install readers (Figure 13-2) at each restricted area entry point. Cards, of course, must be

Figure 13-1. Parking area access controlled by magnetic key card operated gate. Courtesy of Access Control Systems.

Figure 13-2. Programmable magnetic key card reader. Courtesy of Access Control
          Systems.

strictly controlled and issued only to those persons who are authorized to
enter the restricted area being protected. This system provides good control
but has some definite drawbacks also. As with keys, a lost or stolen card
compromises the system and requires that new cards be made and the reader
reset to accept the new cards. The same situation occurs if an employee is
terminated and fails to return his key card. Further, using the single reader
system it is possible to counterfeit the cards with comparative ease, as has
been done to permit entry into parking lots or garages.

A much more sophisticated total access magnetic card system can be
obtained. Such systems are manufactured by several reputable companies.
The systems with which the author is most familiar have been designed to
eliminate the drawbacks normally found in such systems.

Some of the features which may be found in a total entry system which
had been tailored for the needs and physical layout of a particular facility
would be as follows:

1. The key cards may be given their magnetic distribution by the
    user. This eliminates even the manufacturer of the equipment
    from gaining knowledge as to how the system is set to function.

2. As with the banking machine, the system can be designed to require a coded series of numbers to match the card presented before it will grant entry into the restricted area.
3. A printout can be obtained (Figure 13-3) of each card presented to each card reader, indicating the card number. This allows determination of who entered the area, the time and date of entry, the time and date of exit and the identification of the area in which it was used.
4. The system will reject, but record, a card that has not been authorized to enter an area where it is presented. An audio alarm can be installed at the control center to give the alarm anytime an unauthorized card is presented.
5. The system can designate time periods during which employee key cards may be honored and if presented at any other time entry will not be granted. For instance, if a day time employee of the data processing center attempted to return to the computer room late at night, he would be denied entry; his card would be recorded as having made the attempt and if an alarm has been installed, it would be activated.

Figure 13-3. Printer for magnetic key card entry system. Records months, day, hour, minute, card number used and door number that has been entered. Courtesy of Access Control Systems.

6. In the event an employee terminates and fails to return his key card, that card can be immediately denied access and thus rendered useless for any further use unless re-instated.

7. Key cards are adaptable to having a photo identification card incorporated onto the back of the card. Thus, the identification card may be used to gain entry onto the protected property and the magnetic key card used to gain entry into those areas the card holder is permitted to enter.

8. Key cards may be designed to permit entry into a number of restricted areas yet may exclude other restricted areas.

9. Key cards may be tied into data processing banks and utilized to provide attendance data for payroll purposes, thus eliminating the necessity for time clocks and time cards.

10. Cards can be coded to prevent the use of a card by more than one person. For instance, an employee entering through a turnstile cannot hand his card back to another person for his use. Once this card has been used to enter, it will not permit entry until it has been used to leave the area.

11. Key cards also have other uses and with continuing development the sky is probably the limit. At the present time, the key card can be set to control use of copying equipment to eliminate unauthorized and unnecessary use or copying of documents by unauthorized persons. They can be used to control traffic into and out of the protected area. The magnetic card entry systems are a definite asset to your total security program.

There are still other control methods and systems such as the use of data processing equipment to compare fingerprints or handgeometry of persons authorized to enter the protected area. These, coupled with code words and personal recognition, are in use in some of the more highly secured and classified government installations.

It may seem that the electronic age is taking over, yet the effectiveness of the system still depends upon the man observing the monitor, reading the printout or responding to the alarm which makes the program valid. That man, of course, is the security officer.

QUESTIONS FOR DISCUSSION

1. Explain the exchange system of movement control.
2. How is the multiple system of movement control instituted?
3. What are the problems, if any, of using a remotely controlled electrical strike in conjunction with a closed circuit television monitor?

4. Does the use of a single reader, magnetic card system provide the required access control security?
5. What are the advantages of a totally integrated magnetic card entry system?
6. What services, other than entry control, can be provided by a totally integrated key card system?
7. Is it necessary to recover an issued key card to exclude its use in a key card system?
8. What other methods of entry control are available which provide an even greater degree of security?
9. What are the problems encountered in using a separate sub-master key system to control access to a restricted area?
10. Has the security officer's importance declined due to advances in electronic aids to the security system?

MAJOR CONSIDERATIONS

1. A totally integrated magnetic key card entry system is more cost effective than employing the exchange or multiple systems of access control.
2. Lost or stolen key cards can be programmed out of the system and rendered ineffective.
3. The print out provided by the integrated key card system provides a record of all entries and attempted entries.

Chapter 14

CLOSED CIRCUIT TELEVISION

A closed circuit television (CCTV) system, which has been engineered to fit the particular needs of a protected area, is a definite plus to the overall security program. If the system incorporates a microphone and speaker at selected camera locations, it is an even more valuable adjunct.

Have you ever yearned for the luxury of having sufficient manpower to station a man at each weak point in your system as well as in those areas most likely to be the target of pilferers or saboteurs? With closed circuit television you can complete, simultaneously, what otherwise would take a great deal more manpower and time to accomplish.

BENEFITS

Since very few firms, if any, provide unlimited finances to their security budget, the cost of installation of a CCTV system will have to be justified before management will consider the move. Part of the cost will be justified by the increase in security protection and awareness with no corresponding increase, or possibly even a decrease in the number of security personnel required.

Several very beneficial side effects have been recorded by companies following the installation and use of CCTV. For one, a definite decrease was noted in the incidence of industrial accidents which had been caused by gross carelessness, recklessness or as the result of employee horseplay.

The second unforeseen benefit was that production efficiency increased in those areas where cameras had been installed. Critics of surveillance systems might term the installation of CCTV as a "big brother" type invasion of the workers' privacy. However, when a safer work place is being provided and the worker is being given more job security by increasing productivity and reducing losses from accident, theft or sabotage, the majority of the work force should not object. This is particularly true in those companies which provide employee profit sharing.

CAMERA PLACEMENT

Once a decision has been reached to install a CCTV system, recommendations may be requested as to camera placement and perhaps the type of accessory equipment to employ, such as lenses, monitors, all-weather housings or pan and tilt devices.

It should be recommended that a security consultant be called in to survey the protected area and to provide recommendations for locating cameras and monitors. His recommendations should also include the type of accessories which should be used, if any, to increase the value of the system.

Should company management feel that employing a consultant would be an unnecessary outlay of the firm's finances, the risk manager could make the selection of camera locations. After making the site selection, he should contact a reputable local supplier of CCTV equipment and get his recommendations as to the proper equipment and accessories needed to provide the necessary coverage at each selected location.

To assist in determining the best locations for the installation of cameras, the following format is provided:

1. Convene a meeting consisting of representatives from management, security, maintenance, production and shipping/receiving personnel.
2. Discuss and prepare a list of every location within the protected area where the installation of CCTV cameras would be useful.
3. Next to each location on the list write down all the benefits to be derived by the placement of that particular camera.
4. Revise the list to record camera locations in the order of importance to the overall safety and security program.
5. Pare the list of locations by determining those locations where the purchase, installation and maintenance cost of the equipment could not be offset by the usefulness derived from the equipment.
6. Keep in mind while designing the system that the entire system does not have to be installed or implemented at one time. A basic system may be installed in one fiscal year and added to as the budget allows.
7. While in the planning stage, any future expansion requiring construction should also be considered. Even if it is decided that CCTV will probably not be required initially, it is much more economical to install the necessary cable and wiring at the time of construction than at some later date.

At any time that new construction is contemplated, an attempt should always be made to build in as much security protection and hardware as possible. This includes alarm systems, entry control, fire control and alarms, secure doors and locking devices in addition to CCTV.

In selecting CCTV equipment, consider the available light at the selected locations in order to be sure that the camera is able to pick up and identify potential intruders or pilferers. Low light level cameras (Figure 14-1) are available, but if the protective lighting system is adequate there should be little or no problem.

Another important item at selected locations would be the installation of a good zoom lens which allows for a close-up of any suspicious activity.

For cameras which are installed out of doors and exposed to the elements, it is necessary to insure that the camera is encased in an adequate all-weather housing (Figure 14-2). Some cameras should have pan and tilt capabilities so that they will be able to follow suspected movement and to give the camera positions more versatility.

MONITORS

Monitors also come in all sizes, shapes and with varying degrees of capabilities (Figures 14-3 and 14-4). One good type of monitor to consider comes

Figure 14-1.   Low light level camera uses a silicon intensifier target (SIT) tube to provide crisp TV pictures in light levels as low as 0.0001 footcandles of scene illumination. Courtesy of RCA CCTV.

Figure 14-2. Prepackaged camera with integral environmental housing can be obtained with fixed or zoom lens. Courtesy of RCA CCTV.

Figure 14-3. CCTV desk top monitor with 10" screen. Courtesy of RCA CCTV.

Figure 14-4.   CCTV desk top triple monitors with 6" screens. Courtesy of RCA CCTV.

with a 9" screen and is of modular design to permit additional monitors as the system is expanded. Some camera positions may be monitored by one monitor with that position as its only function. However, in most cases two or three camera positions may be controlled by one monitor equipped with an automatic sequencing device to switch from camera to camera. If this system is used, the monitor should be equipped with a manual override to permit the operator to stay on one camera and use the pan/tilt and zoom lens capability to zero in on any suspected activity he may observe.

VIDEO TAPE RECORDERS

There is still another accessory to the CCTV system which will generally prove very valuable in cases of pilferage or intrusion where a record is desired for later positive identification or as evidence of the sequence of events. This, of course, is the video tape recorder (Figures 14-5 to 14-7). With this device connected to a system the operator can, with the touch of a button, have a permanent visual record of what has taken place.

PURCHASE AND INSTALLATION OF EQUIPMENT

When purchasing equipment and contracting for installation, it may be beneficial to request bids from competing businesses. If so, make sure that bids are only accepted from well-known and reputable businesses. Also, before signing

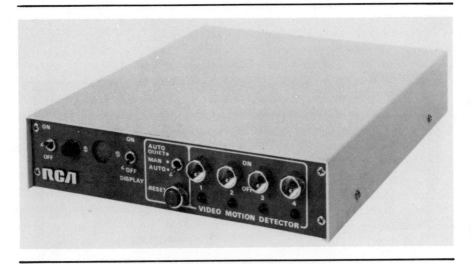

Figure 14-5.   Video motion detector alerts operator to motion in a scene. Can be used
in conjunction with sequential switcher for automatic call up of alerted
camera. Courtesy of RCA CCTV.

Figure 14.6.   Video switcher. Courtesy of RCA CCTV.

Figure 14-7.   Video date time generator puts day, month, year, minute and second on video signal or on taped relay when video tape recording is made. Courtesy of RCA CCTV.

any contractual agreement arrange for a field test of the equipment to be purchased. The test should be conducted at the location where the equipment is to be used, to insure it meets the standards set and that the cameras being purchased are adequate with existing lighting conditions.

Once installation has been made, insure that each officer who is to be assigned to operate the equipment and monitor the cameras is fully trained in the proper use of the equipment either by factory personnel or dealer personnel.

As with alarm systems, the best CCTV system can be of great value to the overall program but it cannot take action or respond to any situation. So, once again, the effectiveness of the system depends upon the security guard force. The operator must remain alert to detect potential problems, and security personnel must be able to respond to the problem and take whatever action is required.

QUESTIONS FOR DISCUSSION

1. What justification can be offered for the expense required to install and maintain a CCTV system?
2. What are some beneficial side effects realized by utilizing CCTV?
3. Will installation of CCTV decrease the number of security personnel required to properly staff the guard force?
4. Who would you contact to receive guidance in the installation and selection of a CCTV system?

5.   How would camera placement positions be decided?
6.   Who should be consulted concerning camera placement? Why?
7.   What advantage can be obtained by using a camera with a zoom lens?
8.   Should monitors be equipped with automatic switching devices, manual switching devices or a combination of the two methods?
9.   What would be gained by having a video tape capability tied to the CCTV system?
10. Will CCTV ever replace the security officer?

MAJOR CONSIDERATIONS

1.   The installation of CCTV provides the capability of constant monitoring of critical areas at less cost than would the utilization of security personnel.
2.   CCTV provides, in addition to its loss prevention function, additional benefits to production and other facilities.
3.   A video tape recorder will provide a record of suspected activity observed on any camera included in the CCTV system.

III. PLANNING:
PROCEDURES AND
EXECUTION

Chapter 15

FIRE PREVENTION METHODS

FIRE!!! Friend or Foe?

FIRE!!! God or Devil's advocate?

Fire! From man's earliest history, mythology of every civilization speaks of fire and man's worship of fire as a god or as the messenger or symbol of God. These beliefs often followed or coincided with the worship of the Sun as a god.

Early man discovered that fire gave him warmth, provided him with light, made his food tastier and provided him with a protective shield against predatory animals. Man also realized that fire could hurt him and cause his death. Man did not really understand fire and because he did not, he attached a moving spirit to fire and made it a god. How man first came into contact with fire is not known. Fire probably came to man as a result of a natural phenomenon such as the eruption of a volcano or a tree that was ignited by a lightning bolt. As man is apt to do, if a fire destroyed the village where he lived, he would select a child or maiden to be sacrificed to the fire god. Fortunately for today's young ladies, we understand a little more about the origin of fires.

Mythology has other answers however. Heroes are believed to have descended into the underworld and to have returned with the fires of hell which they used for their own good and comfort. Throughout the history of man we find references to the underworld, hades, hell or the inferno. These are places of eternal damnation to be feared above all things since it has been said that anyone venturing into these worlds shall be damned eternally to be burned and suffer the pain of hell's fires.

Man's views toward fire have changed little in his centuries on earth. He has found more and more ways to harness the energy provided by fire and has put this energy to work. Picture, if you will, where man would be if it were not for fire to heat his home and workplace; to power the generators which provide his electrical current for light and power, to move the internal combustion engine, the jet engine and the steam engine; to heat his food; or to power his factory. Without fire, man would still be in the stone age.

The main concern of security professionals with fire is in its role as the killer of mankind and the destroyer of property. Man soon learned that fire which is allowed to burn without control spreads faster and becomes too powerful for a single man to control. Early man organized his tribe, village or town to prevent and fight fires. In the early history of the United States, fire guards were appointed to patrol the village each night and to give the alarm if a fire was discovered. Each citizen was required to provide his own fire bucket, fire ladder and long swabs for reaching burning roofs. Every able-bodied person in the village was required to answer the alarm to fight against the spread of fire. In 1648, Peter Stuyvesant, then Director-General of New Netherlands, appointed four men, who were called fire masters, to inspect buildings and enforce fire prevention measures. These men insured that chimneys were kept clean and inspected to see that there was an adequate supply of fire buckets, ladders and swabs.

Today, within protected areas or facilities, the security officer has replaced the night guard. Built-in fire equipment and modern fire departments replace the citizens of the village, yet fire is still one of the greatest threats we face. Consider Chicago and San Francisco; although these fires occurred years ago, they still serve as prime examples of what havoc can be wreaked by the power of fire. Arson, one of the most dangerous and loss producing crimes known to man, is ever increasing in frequency. Fires which are intentionally started with the idea of totally destroying the target building or complex are the result of many motives. The primary motive in most of today's arson fires is profit. Buildings are burned to collect insurance money, to save demolition cost, to hide the evidence of other crimes, out of hatred or revenge and a few are set for the thrill of witnessing the building burn. Some of the latter are the work of pyromaniacs, while others are burned just for the feeling of power in being able to destroy.

The guard force has many responsibilities for the prevention and control of fires. In most situations where guards are employed, one of their primary duties is to patrol the protected area, being alert for fire hazards; to report on, or neutralize; to be available to provide the earliest possible alarm in the event of a fire, and thereafter to take whatever immediate action that is available to them. These actions and their priorities are covered in more detail in the next chapter.

If a guard force is to be effective in preventing, discovering or fighting fires, it will be necessary for them to know the basic elements required to start and maintain a fire; the stages of fire; the classes of fire, as determined by the type of fuel being consumed; and the types of fire fighting equipment that are normally available to assist in the fighting of fires.

ELEMENTS OF FIRE

Fuel

In order to have a fire there must be some form of combustible fuel to consume. Just as man needs food to sustain himself, so fire requires fuel.

Heat

For a fire to start, the temperature of the fuel must be raised to its ignition point. The degree of heat required varies with the type of fuel. The continued burning of the fuel requires that the temperature remain at or above that which is required for ignition.

Oxygen

There must be enough oxygen present to allow the ignition of the fuel and to sustain the burning of the fuel. The amount of oxygen required will vary depending upon the size of the flame or the area of the fire. A match may be ignited in a closed air-tight room by using up the oxygen present in the room before it was closed. However, as the remaining air was consumed, the fire could no longer sustain itself and would go out.

EXTINGUISHING FIRE

It should be remembered that to start or sustain a fire, all three of the elements must be present. In the prevention of fires we attempt to insure that the elements required for ignition of fire do not come together. Once a fire has started we use our knowledge to put out the fire by removing any of the elements it needs to burn.

Fuel

Fires can be prevented by careful storage of the materials or liquids which have a low ignition point. A patrolling guard who notices that gasolines, oils, paints and other such materials have been stored in close proximity to buildings or other highly flammable materials should report this condition as potentially hazardous. In surveying the area to be protected, the supervisor may be able to see many potentially hazardous areas or conditions which if corrected would greatly reduce the possibility of fires, for example, acetylene torches being used in the vicinity of lumber, discarded cartons or packaging materials and spark-producing grinders or other machinery too close to saw dust, paint areas or gasoline pumps. The possibilities are endless, and security personnel should always be on the alert to report and eliminate such hazards.

What about stopping the fire once it has started? Remember that all three elements are required to sustain a fire. Accordingly, fire can be put out by taking away the fuel. The men who fight forest fires are experts at this method. They devise fire breaks or start back fires to take away the fuel and effectively stop the spread of the fire. In the San Francisco fire, as in other widespread city fires of years ago, entire blocks of houses and buildings were blown up to take away potential fuel and thus limit the spread of the fire. On a smaller scale, if a truck were on fire inside a garage, taking the truck out of the garage would not put out the fire but would limit the fuel for the fire to the truck and make it easier to extinguish or control. If the center of a lumberyard caught fire, it might be possible to remove lumber, not yet burning, and thus effectively stop the fire for lack of fuel. As can be seen, the removal of fuel is usually not the fastest or even the most satisfactory method of putting out fires, but there are times when it is the only method that is available.

Heat

U.S. Navy ships, operating in the South Pacific or other tropical waters, used to run a continuous stream of water, during the heat of the day, over decks which covered ammunition storage rooms. This was used as a preventative measure to preclude the possibility of a fire or explosion being brought about by the heat of the sun on the metal decks. In the prevention or fighting of fires, one of the main weapons is keeping the temperature of fuels below their ignition point or, once burning, to lower the temperature below the ignition point of the burning fuel. Normally, water is used to cool off a fire but, as will be shown later, the method used in fighting fires will depend greatly on the class of the fire being fought.

Oxygen

A classic example of the cost of not possessing sufficient knowledge of the elements necessary to sustain a fire was seen in the burglary of an armored truck company vault. This may have otherwise been a perfect crime. A crime can only be perfect when the fact of its commission is known only to the perpetrator. The persons who planned and executed the theft of the money in the vault wanted to make it appear that the contents of the vault had been consumed by fire. They put plastic bags containing gasoline into the vault and used fuses that a fire would consume, thinking that no evidence would be left to determine how the fire started or that a theft had taken place. Having done all this, they then closed the vault doors. The fires ignited as planned but soon consumed all the oxygen in the vault and the fire went out, leaving evidence of the theft as well as the fire.

For students of scientific investigations, it would not have mattered if the fire had consumed the entire contents of the vault since an analysis of the residue would have revealed the use of gasoline in the start of the fire and the volume of ash left as a result of the fire would have indicated that the money had been removed prior to the start of the fire.

Fires are not often fought in airtight rooms in which they can be enclosed. However, through the use of dry chemicals, foam and other materials having smothering properties, fires can be extinguished by eliminating the required oxygen.

STAGES OF FIRE

Fires, excluding those started by an explosion or other method resulting in immediate ignition, go through four distinct states.

The first of these stages is known as the incipient stage. In this stage there is no visible smoke, flame or sufficient heat to be detected. A fire may remain in this stage from minutes to several hours.

The second stage is known as the smouldering or smoke stage. As the fire develops, visible smoke may be detected rising from the burning mass. During this stage there is still no flame and very little detectable heat.

The third stage is known as the flame stage. It is at this point where ignition actually occurs, flames are noticeable and the smoke decreases in volume as the flames and heat increase.

The fourth and final stage is known as the heat, conflagration or explosion stage. This stage closely follows the flame stage, and once this stage is reached the fire can normally be said to be out of control.

CLASSES OF FIRE

As previously stated, fires are classified by the type of fuel that is being burned. This classification is also extended to determine the type of fire extinguisher which should be used to put out the fire. Learning the classes of fire and the symbols that are assigned to each class, will facilitate determining the proper fire extinguisher to use.

Class "A" fires are those that burn fuels consisting of normal combustible materials such as wood, paper, dried grass and cloth. This class of fire has been assigned the symbol "A" in the center of a green triangle. Fires burning these fuels are normally fought by lowering the temperature of the fuel below its ignition point. Water or water-based extinguishants are normally used on this class fire.

Class "B" fires are those that burn fuels consisting of flammable liquids such as cooking grease, paint, gasoline, alcohol and oils. This class of fire has been assigned the symbol "B" in the center of a red square. This type of fire is normally fought by excluding oxygen from the fire through the use of dry chemicals or foam-type extinguishants. It is possible to use water on this class fire to cool it below the ignition point if the water is applied with a fogging-type nozzle. To spray water directly on this class of fire would only spread the fire over a larger area.

Class "C" fires are those that burn in live electrical equipment such as fuse boxes, ovens, motors, switches and various types of electrical appliances. This class of fire has been assigned the symbol "C" in the center of a blue circle. In fighting these fires an extinguishant which will not conduct electricity, such as a dry chemical or carbon dioxide must be used. Water, foam or other water-based extinguishants should never be used on this class of fire.

Class "D" fires are those that burn fuels consisting of flammable metals. This class of fire has been assigned the symbol "D" in the center of a yellow star. Fighting these fires is confined to the use of dry chemicals, powdered graphite and other specially prepared powdered agents.

SECURITY PATROL

Security personnel should be provided with diagrams indicating the location and type of extinguishers in the facility where they are employed, and during patrol rounds they should physically locate the extinguishers available for use.

In most situations, security officers patrolling the area will make weekly checks of extinguishers to insure that they are ready for use and have been recently inspected by qualified maintenance personnel. They should also check to insure that the proper symbol for the extinguisher is displayed above the location, or at least that the symbol is on the extinguisher itself. The various extinguishers available will be discussed with other fire fighting equipment in the next chapter.

Security patrols should also check for good housekeeping procedures to make sure that fires do not have a place in which to spontaneously combust. These checks should be a matter of routine duty which should become second nature to the professional security officer.

Fire fighting is a learned science which requires not only knowledge but also the ability to put that knowledge into effective use. This requires, in addition to a thorough basic training course, a continuing training program consisting of review of necessary knowledge coupled with practical exercises in the use of available fire fighting equipment.

Rarely is anyone, other than the security officer, given the sole responsibility for the protection and safekeeping of industrial or other complexes worth millions of dollars. It has been said before and will be repeated over and over. All of the plans, alarms, fences, lights, locks or other security measures will be of no avail unless the security force performs its duties diligently and faithfully without error.

QUESTIONS FOR DISCUSSION

1. What are the security officer's concerns as pertains to fire?
2. What elements are necessary to start and sustain a fire?
3. How does the knowledge of what is necessary to sustain a fire assist in extinguishing the fire?
4. Name the classes of fires.
5. Why is it important to identify the class of fire before attempting to extinguish the fire?
6. How can knowledge of fire assist the security officer in the prevention of fire?
7. Why is it important to know the type of extinguishant contained in a fire extinguisher before using it on a fire?
8. What are the four stages of fire?
9. How do security patrols assist in the prevention of fires and the reduction of losses from fires?
10. Discuss the benefits man has received from harnessing fire versus the destruction that can be caused by uncontrolled fire.

MAJOR CONSIDERATIONS

1. In learning the elements necessary to ignite or sustain a fire, the security officer gains a tool that will assist in the fighting or prevention of fires.
2. Recognition of the stages of fire can help in determining the type of action to be taken to control the hazard.
3. Being able to identify the class of fire is often critical in safely extinguishing a fire.

Chapter 16

FIRE PROTECTIVE EQUIPMENT

Most, if not all, facilities which require protection will have portable fire extinguishers and other installed or built-in fire fighting equipment available. The security guard force must know how this equipment operates, where it is located and must check it regularly to insure that it is in good working order and is being regularly inspected, serviced and maintained. Normally, it will not be the duty of the security section to maintain or service this equipment, but they should be aware of how frequently that service is required, and in the case of hoses they should insure that they are tested regularly and are in a good state of repair. By conducting at least weekly checks of this equipment, the security guard force can see that the facility management is kept advised on any repairs or replacements that may be required.

FIRE EXTINGUISHERS

Fire extinguishers come in various types and shapes and utilize different methods of expelling the extinguishant. Standards for portable extinguishers are set by the National Fire Protection Association (NFPA), which is a nonprofit technical and educational organization located in Boston, Massachusetts. Materials concerning fire prevention and use of portable extinguishers are available from the NFPA. Testing and other classification of portable extinguishers is conducted by the Underwriters Laboratories, Inc.

Extinguishers are identified by the class of fire for which they were designed and filled. Class "A" extinguishers contain water as the prime extinguishant. The water may be discharged through the use of a pump, the mixture of pressure-producing chemicals activated when the extinguisher is inverted, the use of air under pressure or a gas cartridge contained inside the extinguisher. Instructions are generally contained on the metal plate on the front of the extinguisher. These extinguishers are activated in many different ways, and the security supervisor should insure that he is thoroughly familiar with each type being used in the facility he is protecting in order that he may train his personnel in the use of the various types. Normally the company supplying the extinguishers and/or the local fire department would be more than happy to assist in the training of personnel.

Class "B" extinguishers use three basic agents as extinguishants: carbon dioxide gas, dry chemical (usually bicarbonate of soda which has been treated to preclude caking) and foam. As previously stated, Class "B" fires involve burning liquids. Agents such as vaporized carbon tetrachloride, which were once widely used, have been almost completely discontinued due to their toxic and corrosive effects. If an extinguisher using this agent is found, a management representative should be advised that this is an undesirable and unsafe agent and should be replaced. Activation of the carbon dioxide is usually accomplished by squeezing a trigger device on the extinguisher. A safety pin prevents the accidental discharge of this extinguisher and must be removed prior to its use. Most extinguishers that contain air or gasses under pressure will have a gauge indicating the degree of pressure required for use and will indicate the actual pressure in the extinguisher. Some foam extinguishers, as some water extinguishers, must be inverted to be operated. Some of these begin operation as soon as they are inverted and others must be hit on the ground after inversion to puncture the pressure-producing chemicals. In any event, if instructions do not appear on the front of the extinguisher, it should be replaced with an extinguisher containing an instruction plate.

Class "C" extinguishers use both carbon dioxide gas and dry chemicals to fight fires in electrical apparatus. As a precaution, whenever possible, the power should be cut off prior to attempting to extinguish a fire in any electrical appliance or apparatus. However, since this may not always be possible, it should be determined that the type of extinghishant being used does not conduct electricity and that the nozzles are constructed of plastic, rubber or other nonconductive materials.

Class "D" extinguishers, as discussed, use dry chemicals and powders. This type of fire must be smothered (have the air excluded) to be extinguished.

The entire range of extinguishers which are available has not been covered here. However, from types discussed and from any other types found within the protected area, the security supervisor with the help of the supplier or local fire officials, should be able to initiate training classes for his personnel on the types of extinguishers in use.

In any event, the security officer's first duty upon discovering that a fire is in progress is to call the fire department and report the location and extent of the fire. There should be no exception to this rule. The fire may look small enough to be extinguished through the use of a portable extinguisher and perhaps it is, but that is a chance the security officer may not be permitted to take. Many a building or entire facility has been completely destroyed by fire because it looked like the fire had been put out when it was still smouldering and later broke back into flames which got out of control before the fire department could respond to the call. **This is a cardinal rule: At the first sign of a fire, call the fire department.** After the fire department has been notified, security personnel should insure that the necessary gates are open to admit the fire department and that there will be someone at the gate to direct the department to the fire. If the facility is operating, the alarm necessary to evacuate the affected area should be given. If there is time, then the security force may attempt to extinguish the fire with portable or other installed equipment. Normally, if the plant or facility is operating, fire and emergency teams will be designated and the duties of all involved will be spelled out. If this is not the case, then the security supervisor should recommend that plans be formulated and emergency teams be assigned and trained in their duties.

Wherever possible, extinguishers should be placed in the areas where the class of fire they are designed to protect against is most likely to occur. This will prevent an untrained person, in a moment of panic, using a Class "A" extinguisher on an electrical fire and possibly electrocuting himself. An example would be having only foam extinguishers in paint storage or gasoline dispersal areas or having only carbon dioxide extinguishers present in the electrical generator room. Here again, the security force may not be consulted before the extinguishers are located; however, if security personnel note an extinguisher which should be relocated, this should be reported to management personnel.

AUTOMATIC SPRINKLER SYSTEMS

Automatic sprinkler systems are also found in many facilities which require security protection. Although they are designed to be automatic, each member of the security guard force must have a working knowledge of the type or types of system(s) installed in the facility. Basically, there are three types of automatic sprinkler systems which are designed for entirely different purposes. It is possible that all three types may be found in a single facility. These three systems are known as wet pipe, dry pipe and deluge systems.

The wet pipe system, in order to operate, must have a supply of water throughout the entire system of pipes. Once the sprinkler head has been ruptured, the water is immediately discharged in the area of the rupture.

The dry pipe system uses compressed air to hold the water at the riser location until a sprinkler head has ruptured. The escaping air then allows valves to operate and permits water to flow into the system. This type of system is best employed in areas where if water remained in the pipes, extreme cold might rupture the pipes and wipe out the system.

The deluge system, as with the dry pipe system, has no water in the pipes until a fire has started. The sprinkler heads used with this system are very similar to small nozzles which are designed to direct the flow of water to a specified location. The water is held at the riser location and allowed to enter the system by use of either manual or automatic control. If automatic, the system is used in conjunction with heat sensing devices or other fire detecting devices which operate the riser valves. Once water enters the system, it flows out of the nozzle-like heads which are always open.

Security personnel must have written instructions on their duties in connection with insuring that the sprinkler system installed on the protected property is working properly. If they must check riser valves to insure that they are in the proper position or if they must monitor pressure gauges, their training must be thorough and complete and their instructions explicitly reduced to writing. The security supervisor must never permit one of his personnel to work alone within the protected facility unless he is confident the officer is fully trained and capable of handling any situation which may arise.

STANDPIPE HOSES

Yet another system which may be installed to help protect against a potential fire is the standpipe hose system, which is installed and operated separately from the sprinkler system. This system is not routed through the riser valves controlling the sprinkler system to insure that neither system will rob the other of the pressure necessary to operate effectively.

When the system of standpipe hoses is installed in a multi-story building, the pipe that provides the water for the system is generally run up the wall of the fire well which houses the escape stairs. The hoses are located on each floor near the fire escape door from that floor. The designed location was chosen in order that personnel fighting the fire on each level would have a clear and safe escape route down the fire escape stairs.

Hoses to this system should be maintained in cabinets, designed for that purpose, wherever possible. Hoses which are left in the open are often unnecessarily exposed to caustic substances which causes their deterioration long before their usefulness should have ended. This is an unnecessary expense and also poses a threat to life and property in the event a faulty hose should be relied upon to extinguish a fire. Plastic covers which are generally red, are available and are appropriately labeled for use on hoses that are stored on reels or horizontal brackets.

Hoses should always be connected to the standpipe in order that they may be available for immediate use in the event of a fire. Additional couplings may be available for the use of the fire department or company-appointed emergency team personnel.

Hoses may be equipped with an appropriate nozzle, selected for the type of fire most likely to be encountered in the area of each particular hose. Nozzles are available from a straight heavy stream to the fog spray nozzle. Some nozzles presently on the market may be adjusted to any type of water spray that is desired.

The operation of a standpipe hose normally requires two persons: one to lay out the hose and the other to guide it off the reel or other holding device. Caution should be exercised to insure that the hose is fully extended prior to releasing the water into the hose. Water released prior to the hose being fully extended will cause the hose to whip violently and possibly injure the personnel using it. Once the hose has been extended and the water is flowing, both men should handle the hose and direct the water to the foot of the blaze.

Whatever method of fighting the fire has been brought into play, i.e., portable extinguishers, standpipe hoses, sprinkler systems, or fire department equipment, caution must be exercised to limit the water damage to an absolute minimum. You *must* insure that sufficient water has been used to completely extinguish the fire. Personnel handling hoses should remain in position to play more water on the burning area should the fire rekindle.

HOSE HOUSES

Whenever hoses and hydrants are maintained outside the protected facility, the house should be constructed so that it has shelves for the hoses and can accommodate the hydrant and other equipment. If at all possible, the house should be so constructed to allow the fire hydrant to be as close as possible to the front or open side of the house. When the guard force inspects these houses, they should check the hoses for cleanliness and serviceability and make sure the other equipment is present and in good condition. Where possible, hose houses should be numbered and a corresponding number used to identify equipment assigned to that particular house. NFPA has listed what they believe to be the minimum amount of equipment required for an industrial location hose house. This list may be modified to fit the needs of a particular area.

- 2 underwriter play pipes
- 1 pair play-pipe brackets
- 1 fire axe
- 1 pair fire-axe brackets

- 1 crowbar
- 1 pair crowbar brackets
- 1 extra hydrant wrench (in addition to wrench on the hydrant)
- 4 coupling spanners
- 2 hose and ladder straps
- 1 underwriter play pipe holder
- 2 2½-inch hose washers (spares)

In the event the hose houses cannot be located over the hydrant, they should be located as close as possible to the hydrant and maintained as described above. Houses should be kept under lock or seal. If a lock is used, it should be the break away type. Any method can be used which will insure the integrity of the hose house while at the same time providing the swift access required in an emergency situation. The hose houses should be regularly inspected by the security force and the condition and inventory of each house reported to management. Consideration should be given to typing all equipment contained in each house and keeping it on the inside door so that the taking of an inventory will be greatly facilitated.

FIRE DOORS

It would not normally be the responsibility of security guard personnel to conduct inspections and/or maintain fire doors. However, each member of the guard force must know the location and operation of such doors and how their being closed affects movement from one point to another within the protected facility.

During the initial and ongoing training sessions, mentioned in the preceding chapter, the supervisor should provide his guard force with information as to type and location of any hazardous chemicals, liquids or other volatile material being stored or used within the protected property. In the event these materials require special extinguishants or special handling, as in the case of poisonous gases being given off, this information should also be given to members of the security force.

The supervisor should be constantly asking himself if the personnel under his control have been provided with the training and information they need to accomplish their assigned duties.

QUESTIONS FOR DISCUSSION

1. Where can information be obtained concerning the type of fire fighting equipment that is available?

2. Why is it important for the guard force to know the different types of fire extinguishers and how they are activated?
3. Name some of the methods which are employed to activate portable fire extinguishers.
4. What is the first thing a security officer must do upon discovery of a fire? Are there ever any exceptions? Why?
5. Name the types of sprinkler systems which are used to protect against the spread of fire.
6. What is the advantage of using a dry system?
7. Where should hose houses be located and how should they be secured?
8. Where can you expect to obtain assistance in training your security force in the operation of fire fighting equipment?
9. Where would you normally find a stand pipe and hose installed in a high-rise building?
10. Discuss the methods used to employ a stand pipe hose against a fire.

MAJOR CONSIDERATIONS

1. The first action to be taken upon discovery of a fire is to notify the fire department.
2. Fire fighting equipment must be located in the proper place and must be inspected regularly.
3. The three types of sprinkler systems, wet, dry, and deluge, each have different applications.

Chapter 17

SAFETY PROCEDURES

Modern technology has made life easier and more satisfying for us all. It has, for the most part, taken the manual out of manual labor by providing us with a variety of labor-saving devices, machines, vehicles and procedures. For all of the advantages these technological advances have given us, they have the capacity to do us harm. Added to all the old hazards we face throughout our daily routine, it seems we are taking a chance on being involved in an accident just by getting out of bed each day.

For the security force, the responsibility for safety does not end with taking steps to prevent their own involvement in accidents. They are also required to insure the safety of all persons who enter the area which they are assigned to protect, as well as to prevent accidents to the property therein.

In the past the disciplines of safety and security, although closely aligned, have been distinctly separated. Today we find that they are often combined under the supervision of a company executive who is charged with responsibility to oversee all aspects of loss prevention.

As pointed out in Chapter 6, these positions have come about with the realization that any activity which has the capacity for lost revenue, and thus profits, needs to be given the same management attention as marketing, manufacturing, financing and all other aspects of operating a successful business. Safety and security professionals filling these positions are being redesignated as risk managers or loss prevention managers to denote their added responsibilities and importance to the continued profitability of the business.

SAFETY AND PROFITABILITY

One of the major areas affecting the profitability of a business enterprise today is safety. The lack of a good workable safety program can rob a company of its profits just as quickly and certainly as an armed band of criminals carrying out the profits in canvas bags.

This loss begins with the employer's payment of insurance premiums to cover job-related injuries and accident-damaged/destroyed equipment or structures. These premium costs are directly tied to the frequency and extent of claims for such injuries and damages. A good safety and loss prevention program can result in the savings of millions of dollars in premium cost alone.

As if this were not enough, the losses of revenue from accidental causes are just beginning. Injured workmen and damaged/destroyed property also cause a large loss of revenue from loss of production and productive capability. While workmen are recuperating from their injuries, equipment is being repaired or replaced and structures are being rebuilt, workers are idle, causing the employer's share of unemployment insurance compensation to escalate. Seriously injured or dead workmen will have to be replaced, necessitating extra cost to the business for recruitment, employment processing and training of the replacement workers.

Probably the greatest loss incurred by firms that manufacture and market a product is the loss of a market for that product. They have been selling to customers at wholesale and retail levels partly because of price, availability, brand name identification or reputation for reliability. Whatever the reason that they control a share of the market for a particular product, when they are unable to deliver, their customers must go elsewhere to procure a similar item from a competitive manufacturer. When they are back in production they may have to charge more for their product and the market loss which they suffer may prove to be irreversible.

As a case in point, a plant operated by a national chemical corporation to manufacture sponges, used in both domestic and industrial settings, accidently caught fire and was destroyed. Prior to the fire this product accounted for the lion's share of the market. In the absence of this product from the market, while rebuilding was taking place, the market was lost to competitors. With the reduced demand for its product this plant was unable to return to the quantity of production it had maintained prior to the accident.

The risk of all these potential loss situations can be greatly reduced or eliminated through the planning, implementation and execution of a viable safety program.

SAFETY PROGRAM

Regulations

In planning a safety program, a good place to start is by studying the regulations promulgated by the state Occupational, Safety and Health Division and those of the federal office administering compliance with the

Occupational, Safety and Health Act of 1970. Add to this the principles of good housekeeping and maintenance, and sprinkle generously with common sense. If a safety plan is to be implemented, to be sure that it is effectively executed means starting by involving everyone that will be affected by the plan and finding ways to motivate all concerned.

This motivation and involvement must begin with the establishment of rules and regulations by management which must be complied with by all personnel, management as well as labor, with enforcement responsibility given to all supervisory personnel of the facility but particularly to the security force. The rules must address themselves to the conduct of employees while at their workplace; the wearing of safety devices and clothing; maintenance schedules for equipment; observance of good housekeeping procedures; safe operation of machinery, vehicles and devices used in the conduct of company business; and disciplinary action for violators as well as establishing procedures for reporting safety violations or observed hazards. Once prepared, these rules should be incorporated into the general work rules for employees.

Training

Training sessions must be initiated for all employees, emphasizing the importance of safety to the individual, not only for his own physical safety but also how safety relates to the stability of the company and in turn to the employee's own job stability and security. Once the employee's education and training on the importance of safety has been accomplished, the next step in implementing a safety program is to provide for a constant and total awareness of the importance of initiating and maintaining good safety habits.

This can be accomplished by follow-up training sessions, insuring that foremen, supervisors and managers continually stress the importance of safety in their contact with employees. Other methods that may be used to obtain the necessary awareness of the importance of maintaining good safety habits are providing awards for good safety practices and records, using signs indicating the number of disabling or productive loss causing accidents, placing safety posters throughout the protected area and establishing rewards or awards for suggestions from employees resulting in greater safety in business operations.

Role of the Guard Force

Safety is everybody's business and for a safety program to be effective, everybody must become actively aware and involved in keeping the workplace safe. Although this total involvement must be achieved, the bulk of the effort aimed at discovery, corrective action and compliance with safety regulations will be the responsibility of the security guard force.

The guard force, after all, are the professionals. They are specifically trained to observe, correct or report hazards encountered. They are the group most familiar with all areas within the perimeter of the facility and they are on duty in the facility 24 hours per day, 7 days per week. This is why they have been appointed by management as the enforcers of the company's rules and regulations.

Each member of the security force must be alert at all times for potentially hazardous conditions and/or situations. It does not matter if the officer is on patrol, en route to the cafeteria for a break, reporting for work or leaving the workplace following completion of his shift; any violation or condition constituting a hazard which is observed must result in affirmative action, on the part of the officer, to eliminate or neutralize the hazard or condition.

Some examples of appropriate action which members of the security force should take are listed below:

1. Observed safety violation by employee
   A. Identify employee, advise him of the violation and require that the violation be discontinued.
   B. Report the violation to the employee's immediate supervisor.
   C. Record the time, place and type of violation, name of violator and action taken for inclusion in safety patrol or special report, directed to facility management, for follow-up at that level.
2. Discovered or observed safety hazard
   A. Correct condition on the spot, if possible.
   B. Where immediate corrective action is impossible, mark the condition as a hazard and report it to the supervisor who is responsible for correcting the condition.
   C. Record the condition, its location and the action taken for inclusion in the risk manager's report to management.
   D. Allow sufficient time for the condition to be corrected, or for corrective action to have been instituted, and check on the progress of the action.

When preparing a training course for the security guard force, which is designed to prepare them for their safety related duties, all areas, conditions and hazards with which the security force will be concerned should be included. Some hazards which might be present in any business facility are listed below in categories to assist in their recognition.

1. Floors, aisles, stairs and walkways
   A. Oil spills or other slippery substances which might result in an injury-producing fall.

B. Litter, obscuring hazards such as electrical floor plugs, projecting material, or material which might contribute to the fueling of a fire.

C. Electrical wire, cable, pipes, or other objects, crossing aisles which are not clearly marked or properly covered.

D. Stairways which are too steep, have no non-skid floor covering, inadequate or non-existent railings, or those which are in a poor state of repair.

E. Overhead walkways which have inadequate railings, are not covered with non-skid material, or which are in a poor state of repair.

F. Walks and aisles which are exposed to the elements and have not been cleared of snow or ice, which are slippery when wet or which are in a poor state of repair.

2. Doors and emergency exits

A. Doors that are ill fitting, stick and which might cause a slow down during emergency evacuation.

B. Panic-type hardware which is inoperative or in a poor state of repair.

C. Doors which have been designated for emergency exit but which are locked and not equipped with panic-type hardware.

D. Doors which have been designated for emergency exit but which are blocked by equipment or by debris.

E. Missing or burned out emergency exit lights.

F. Non-existent or poorly marked routes leading to emergency exit doors.

3. Flammable and other dangerous materials

A. Flammable gasses and liquids which are uncontrolled in areas where they might constitute a serious threat.

B. Radioactive material not properly stored or handled.

C. Paint or painting areas which are not properly secured or which are in areas that are poorly ventilated.

D. Gasoline pumping areas located dangerously close to operations which are spark producing or in which open flame is being used.

4. Protective equipment or clothing

A. Workmen in areas where toxic fumes are present who are not equipped with or who are not using respiratory protective apparatus.

B. Workmen involved in welding, drilling, sawing and other eye endangering occupations who have not been provided or who are not wearing protective eye covering.

    C. Workmen in areas requiring the wearing of protective clothing, due to exposure to radiation or toxic chemicals, who are not using such protection.

    D. Workmen engaged in the movement of heavy equipment or materials who are not wearing protective footwear.

    E. Workmen who require prescription eyeglasses who are not provided or are not wearing safety lenses.

5. Vehicle operation and parking

    A. Forklifts which are not equipped with audible and visual warning devices when backing.

    B. Trucks which are not provided with a guide when backing into a dock or which are not properly chocked while parked.

    C. Speed violations by cars, trucks, lifts and other vehicles being operated within the protected area.

    D. Vehicles which are operated with broken, insufficient or non-existent lights during the hours of darkness.

    E. Vehicles which constitute a hazard due to poor maintenance procedures on brakes and other safety-related equipment.

    F. Vehicles which are parked in fire lanes, blocking fire lanes or blocking emergency exits.

6. Machinery maintenance and operation

    A. Frayed electrical wiring which might result in a short circuit or malfunction of the equipment.

    B. Workers who operate presses, work near or on belts, conveyors and other moving equipment who are wearing loose fitting clothing which might be caught and drag them into the equipment.

    C. Presses and other dangerous machinery which are not equipped with the required hand guards or with automatic shut off devices or dead man controls.

7. Fire protective equipment and operation. (See Chapter 15, Fire Prevention, and Chapter 16, Fire Protective Equipment.)

8. Welding and other flame- or spark-producing equipment

    A. Welding torches and spark-producing equipment being used near flammable liquid or gas storage areas or being used in the vicinity where such products are dispensed or are part of the productive process.

    B. The use of flame- or spark-producing equipment near wood shavings, oily machinery or where they might damage electrical wiring.

9. Miscellaneous hazards

    A. Medical and first aid supplies not properly stored, marked or maintained.

B. Color coding of hazardous areas or materials not being accomplished or which is not uniform.

C. Broken or unsafe equipment and machinery not being properly tagged with a warning of its condition.

D. Electrical boxes and wiring not properly inspected or maintained, permitting them to become a hazard.

E. Emergency evacuation routes and staging areas not properly marked or identified.

As can be seen by these few examples, a security officer should never have time to become bored, and even if the total extent of his duties was only to check and to report on safety hazards, he should never be able to honestly submit a negative report or one containing only "all secure."

Security orders should provide, in addition to the normal observations of security personnel on routine patrol, that at least one patrol on each shift be devoted entirely to the observation and reporting of safety violations or to the discovery and neutralization of safety hazards. The risk manager may want to utilize a special report form which can be used by patrolling officers, on their safety patrol, to report all safety violations or hazards which they observe.

A sample report form is provided (Figure 17-1) which can be changed, added to or have parts deleted, depending on a particular business firm's physical layout and types of hazards normally encountered. This form, with minor changes, should be adaptable for use by a safety patrol within most facilities which utilize security personnel.

QUESTIONS FOR DISCUSSION

1. What are some reasons for implementing a safety program?
2. What effect could a poor safety program have on a company's profitability?
3. How does the loss of productivity contribute to the loss of product market?
4. Discuss the responsibility of the security force for facility safety programs.
5. Why should rules and regulations pertaining to safety be necessary?
6. Discuss the problems which might arise in the storage and use of flammables and other dangerous materials.
7. What are some ways to motivate employees to participate in the established safety program?
8. Name some of the hazards which might be reduced through the issuance and use of protective clothing and equipment.
9. Discuss the necessity for unrestricted exit through doors designated and marked as emergency exit portals.

(FIRM LETTERHEAD)

SECURITY DEPARTMENT
SAFETY PATROL REPORT

I. CATEGORY OF HAZARD OR VIOLATION:

(Circle appropriate item numbers.)

1. Lighting.

2. Floors, Aisles, Stairs & Walkways.

3. Doors & Emergency Exits.

4. Flammable or Dangerous Materials.

5. Protective Equipment or Clothing.

6. Vehicle, Operation & Parking.

7. Electrical Wiring & Equipment

8. Miscellaneous Hazards.

II. VIOLATION OR DEFICIENCY:

(Identify by category number, nature of problem and specific location.)

_____

_____

_____

_____

III. PERSONNEL INVOLVED AND/OR NOTIFIED:

(Specify and fully identify.)

_____

_____

_____

_____

IV. ACTION TAKEN OR RECOMMENDED:

_____

_____

_____

Time Of Patrol _____   Security Officer _____

Date _____   Supervisor _____

NOTE: Copies of this report should be distributed to the Maintenance Department, Department Heads of any department where a violation was noted or whose personnel were involved, to the Risk Manager and to the security force report file.

Initially, when safety patrols are first instituted, the volume of violations and observed hazards may require a more lengthy report form. However, once the majority of the hazards have been identified and corrected a form of this size should suffice for the safety patrol report.

Figure 17-1.   Sample form for reporting safety violation or hazards.

10. How is the security force specially suited for monitoring and enforcing safety rules and regulations?

MAJOR CONSIDERATIONS

1. Safety is everyone's business and everyone must become involved if a safety program is to be effective.
2. Security personnel must be well trained and closely supervised to insure that they recognize and report safety hazards and violations.
3. The establishment of a safety patrol to be conducted on each shift will greatly contribute to the overall effectiveness of the safety program.

Chapter 18

EMERGENCY PLANS AND PROCEDURES

A well-planned security system must include plans and procedures to allow for the efficient handling of any possible emergency situation which might arise. This would include fires, explosions, floods, earthquakes, hurricanes or tornados, volcanic eruptions, riots and other civil disturbances or any life-threatening condition that would require response by emergency teams and evacuation of employees.

These plans must provide for the designation of key personnel who will be authorized to make decisions as to when an emergency exists and to implement the appropriate plan of action. The plan must include the emergency evacuation of the affected area, a method of identifying personnel, formation of emergency teams, establishment of a communications capability, coordination with outside agencies, availability and location of medical services, authority to shut down the facility, procedures which will be followed, and training of all personnel in the duties or actions they are to take in the event an emergency situation is declared.

The chief executive officer, facility or plant manager, whatever the designation, the person in overall charge of operations within the facility will also be primarily responsible for the formulation and implementation of the emergency procedures plan. It will be necessary for this individual to delegate authority to implement the plan in the event an emergency arises during his absence. It will also be his responsibility to designate key personnel and to assign specific duties to each. For the most part the personnel and their designated duties will follow the organizational structure of the company.

EMERGENCY EVACUATION

The purpose of the evacuation of personnel from a facility is to remove them from a potentially dangerous area or structure to a place of safety. The evacuation also gives members of emergency teams unobstructed access to the area affected by the cause of the emergency. During an actual emergency tension runs high and some personnel may be prone to panic. This potential for panic can be greatly reduced if all personnel are thoroughly familiar with the evacuation plan and have confidence in its ability to see them safely out of the danger area.

To achieve this will require many factors, beginning with the designation of routes which will be used for evacuation. It will be necessary to obtain a plot plan of the facility which indicates the location of all departments and operations. It will be necessary to record the number of personnel employed in each department or area, broken down by the shifts when they are present. The plot plan must also show all aisles throughout the facility which can be used for emergency evacuation routes, and it must indicate the location of each emergency exit from the facility.

An analysis must be made of the number of persons who can be safely evacuated through each emergency exit, the time required to move personnel of a given department to the nearest emergency exit and how many of the facility's personnel would be required to leave through each of the designated exits. Following this analysis it may be necessary to establish additional exits to provide for the safe evacuation of all personnel. Once these factors have been established, specific routes and exits can be designated to be used in the event of the necessity of evacuating any given department or section within the facility. The selected routes must be capable of moving the maximum number of personnel present in the facility at one time, should that eventuality arise.

The selected routes and designated exits should be marked on the plot plan, using a color coding method, with a different color being assigned for each department or section to be evacuated. If a color coded identification system has been adopted utilizing different colors for each work unit, the same color codes can be applied to the emergency evacuation plan. The completed plot plan should be reproduced in sufficient quantity to permit its posting throughout the facility and its use in the indoctrination of all personnel.

The next step to be taken is to provide for the safe evacuation of personnel by marking the actual routes using the same color coding used on the plot plan. There are several methods which can be used and are adaptable to any facility which might be encountered. The three most common methods presently being used are

    1.   Printed signs bearing the words "Emergency Exit" and having an arrow pointing in the direction of travel. These signs would be posted at any turning point along the route. This method is particularly adaptable for office areas where the floors are carpeted and which would not lend themselves to the other methods described below.

2. Color coded arrows painted on structural support beams along the evacuation route with the words "Emergency Exit" beneath the arrow. These arrows should be closely spaced and must be marked at any location where a directional change is required. This method is normally used in large open areas such as warehouses and storage areas.

3. A continuous stripe, in the appropriate color, painted on the floor starting at the department designated by each color and continuing to the emergency exit that is to be used by that department. This method is most often found in use in plant production areas.

It will not be sufficient to merely move your personnel out of the area or structure being threatened. They must be moved to a staging area which is located a safe distance from the facility so that any possible explosion or fire would not harm the assembled personnel. These staging areas must be clearly marked using the same color coding as was used in marking the evacuation routes. All personnel not assigned to an emergency team will be required to report to the staging area so that a head count may be made, by supervisory personnel, for the purpose of determining that all personnel have been safely evacuated.

Personnel who have been assigned to an emergency team will report to a central pre-designated location where they will join their team and pick up the equipment and the supplies required to accomplish the duties that have been assigned to their team.

A method or means must be established for alerting all employees that they must leave their workplace and evacuate the facility. This will normally take the form of a siren or bell accompanied by flashing emergency lights. Whatever form is used, it should be electrically operated with an automatic switching device to back-up battery power in the event of a power failure. All evacuation routes should be provided with sufficient emergency lighting to permit the safe evacuation of all personnel during a power failure. This emergency power source should be extended to all lights used to mark the location of emergency exits.

Once the evacuation plan has been completed, routes have been designated and marked, and exit doors and staging areas assigned, consideration must be given to identifying and designating alternate routes. Should primary evacuation routes be blocked by fire, structural collapse or explosion, personnel affected will have to be pre-assigned to alternate evacuation routes. This may be accomplished by assigning each department a secondary color for the route to follow should it become necessary. The alternate routes should be close to the primary route and will have to be capable of handling the additional personnel.

At this point training sessions should be initiated within each department or evacuation unit to fully familiarize all personnel with the evacuation plan. Initially, drills should be held by individual departments walking through their evacuation route and their alternate route. When this has been accomplished it would be desirable to hold an actual drill of the entire facility and system. This can be most economically accomplished at the end of a shift. Employees can be released directly from their staging area to clock out for the day. These drills should then be scheduled periodically, with the frequency determined by the percentage of employee turnover.

EMERGENCY TEAM

The emergency team will be made up of personnel from each shift who possess skills necessary to complete a plant shutdown, shutting off or re-establishing the power supply; contacting and coordinating outside assistance, such as the bomb squad, fire or police department or medical triage teams; cutting off or restoring the water supply; fire fighting; and administering first aid and any other skill which might be required during an emergency situation. An example of how these teams are normally composed and how responsibilities would be divided is listed below.

Team Leader

Maintenance supervisor, facility engineer or whatever the designation, the person responsible for the mechanical operation of the facility will be in charge of the emergency team. This individual is selected because he is usually more familiar with the power sources, facility layout and mechanical operations than any other person.

The team leader will receive his instructions from the facility manager or his designated representative. Based on decisions made by top management, he will coordinate all activities to be conducted by the team. Although this supervisor is a permanent member of the team, he must have an alternate on each shift in the event he is unable to respond to the emergency. These alternates will normally be shift supervisors from the team leader's department.

Assistant Team Leader

The facility security or risk manager is the next permanent member of the emergency team. He too will be familiar with the layout of the entire facility but probably will not be familiar with the mechanical aspects. He will receive his briefings and orders from the team leader.

The assistant team leader must also have designated alternates on the emergency team who will normally be the shift supervisors of the security force. It will be the responsibility of this member to make and coordinate all requests for outside assistance. He will deploy the security force to prevent any unauthorized entry to the area during the emergency by potential pilferers or looters. During such emergencies the security force's communications network will be used to supplement other emergency communications systems.

Technicians

Technicians assigned to the team will consist of electricians, mechanics, maintenance personnel, medics, security officers and coordinators from each department within the facility. The team designated as the facility fire brigade will normally form separately and stand by for instructions from the team leader.

Station assignments and specific duties of each technician will be spelled out in the emergency plan. However, they are not to take any action without clearance from their team leader. Most of these technicians will be involved in the shut down of facility operations should that become necessary. Medical personnel would normally be assigned to an emergency medical station and their primary purpose is to treat members of the emergency team who are injured while carrying out their mission. A security officer will normally be assigned to the team leader as a runner and coordinator for the security manager.

PUBLIC INFORMATION

Any information concerning the emergency that is to be released to the news media or the general public must be cleared through the industrial relations manager, or in his absence the personnel manager or the security manager. The emergency plan should provide that no information is to be released without the express approval of the facility manager or his designated representative.

Notification of key personnel of a facility, in the event of an emergency, will normally be the responsibility of the security force shift supervisor after being advised that such notifications are to be made by the risk manager or the industrial relations manager.

EMERGENCY EQUIPMENT LOCATION

A complete listing of all emergency equipment should be made and attached to the emergency plan. The location of each item of equipment should be exact, and provisions must be made for frequent periodic inspections and necessary maintenance of this equipment.

TRAINING

The need for employees to be trained in emergency evacuation procedures, routes and staging areas has already been discussed. It is also necessary for special training to be given to all persons who have been designated as members of the emergency teams. As mentioned in a previous chapter, the fire brigade will normally be able to obtain their training from the local fire department. Medical personnel, if not registered nurses or trained paramedics, should receive advance training in first aid treatment, including CPR, from the American Red Cross or other qualified agencies or sources.

All members of the emergency team should be given orientation tours throughout the facility to make sure that they are sufficiently familiar with the layout to be effective during an emergency situation. During an actual emergency, the security force will be called upon to perform their normal duties of providing protection to the facility. Additionally, they will be expected to provide such services as controlling traffic exiting the facility and making sure that routes are kept open for the entry into the facility of emergency vehicles such as fire, police or ambulances. They also may be called upon to control crowds which gather near the facility. Security officers who are off duty may be called in to augment the working shift and to provide for extra patrols around the perimeter of the facility.

The uninitiated, observing a security officer checking the identification card of an employee, patrolling the perimeter barrier, making clock rounds or directing traffic during shift changes, might well think that the job was a snap and wish that they could change places and have the lush easy job of the security officer. What they do not see is the midnight patrol of the perimeter barrier with the rain or snow making the task more miserable than lush. They do not consider the times when the security officer is called upon to put his life on the line during a fire or explosion. They do not see the weight of responsibility that is placed upon the security officer who is charged with the protection of life and property within the protected area. They do not understand that without this man and his proper performance of duty there may not be a facility for them to earn their livelihood.

As with most authority figures, the security officer is often maligned and criticized, both for the action he has taken and for action he deemed it unnecessary to take. However, when an emergency arises and their lives may be endangered, these same people will be the first to look to the security officer for protection.

A security force is put to the test daily but never to the extent that they face during an emergency situation. It is at this time that the professionalism, dedication and training of a guard force will decide if people will live or die and if property will be saved or lost. Some of their duties will get boring and they will have to be reminded that the training and strict devotion to duty must be maintained and built upon if they are to be effective when the emergency situation arises.

QUESTIONS FOR DISCUSSION

1. Discuss the method used, if any, to mark the emergency evacuation routes in your facility and how they can be improved.
2. Who is in charge of the emergency team? Why?
3. What official is responsible for the notification and requests for outside assistance?
4. What are some considerations which must be made when planning and implementing an emergency evacuation plan?
5. How important is the role of the security officer during emergency situations?
6. What is the necessity for establishing a staging area where employees can gather once they have evacuated the facility?
7. What is the purpose in the formation and use of emergency teams?
8. In what manner are employees alerted to the necessity for the evacuation of the facility?
9. Discuss the duties of technicians who are assigned to the emergency team such as electricians, mechanics, maintenance and medical specialists.
10. Discuss the benefits of extending the color coding system used on personnel identification cards to the marking of evacuation routes and staging areas.

MAJOR CONSIDERATIONS

1. Emergency plans must be made to allow for the orderly evacuation of a facility to minimize personal injury and loss of life as well as to bring maximum effort to bear on the problems brought on by the emergency, thus minimizing property losses.
2. Training and practical application of emergency procedures is essential to the successful implementation of the emergency plan.

3. The security force is assigned extremely critical responsibilities in times of emergencies and the only way they will be able to successfully accomplish these tasks is through prior indoctrination and training.

Chapter 19

First Aid

The definition most often given to first aid is that it is emergency treatment to an injured, drowning, unconscious or suddenly ill person, given before and until professional medical help can be obtained.

In emergency situations resulting in injuries we instinctively turn to certain professions, other than medical, and expect them to take charge and render the necessary first aid. These professions are policemen, firemen, and safety-related occupations such as plant safety engineers, lifeguards and safety patrols. Within the area he is hired to protect, the security officer is a combination of policeman, fireman, and safety regulator. As such, the security officer must be well versed in all aspects of first aid and must be ready and capable of providing emergency treatment.

Many modern plants and businesses provide first aid rooms which are usually staffed with a registered nurse. They also have, or should have, written policy to follow in case of a medical emergency. These instructions should include the ambulance service that is to be called, what doctor is used by the company, to which hospital the injured or ill person should be taken and the name of the insurance carrier or governmental agency that insures the company against work injuries.

Because security officers patrol the protected area, are located in strategic locations, often have a mode of transportation and are generally the first to be notified in case of emergency, they should be well versed in giving first aid, even if the facility has a nurse on 24-hour duty.

In this chapter basic emergency procedures will be outlined, including priorities of treatment should members of the security force be faced with someone who has sustained multiple injuries. It is strongly suggested that security supervisors attend Red Cross or other comparable training to qualify themselves as instructors in order that they may certify the remainder of the security force. If time and money permit, each member of the security force should be afforded the opportunity of attending a basic first aid class. Even in situations where the supervisor is a qualified instructor, the Red Cross and other agencies giving first aid training will be able to provide better training due to their experience and the resources they have at their disposal. Review and refresher courses in this subject should be scheduled at six-month intervals if at all possible.

The information and techniques described in this chapter were up to date at the time of this writing. However, due to new or better techniques being constantly developed or improved, this or any other book describing first aid procedure should be checked with the American Red Cross to insure that the techniques or procedures described are still considered the best treatment.

There are basically five life and death emergencies which security personnel should be able to recognize and handle. This knowledge could save a life or, conversely, if there is no one present with an adequate knowledge of first aid, it could cost a life. The five emergency situations are as follows:

1. Breathing stoppage
2. Heart stoppage
3. Bleeding and hemorrhage
4. Poisoning
5. Shock

In being faced with a situation which requires giving first aid to a person who has multiple injuries, the treatment which must take precedence over the rest is the treatment of breathing stoppage or heart stoppage. Regardless of the cause of stoppage, every second counts in restoring the ability to breathe or continuing the flow of blood through the heart.

BREATHING STOPPAGE

Keep in mind that a person who has stopped breathing may die within three minutes without positive action to restore his ability to breathe. Action must be taken rapidly and effectively when the victim's breathing has been stopped as a result of drowning, electrical shock, choking or by any other cause.

The mouth-to-mouth method is now considered the most effective way of providing artificial respiration. The procedures which are outlined below should be learned by all members of the security force.

1. Explore the mouth for any obstruction, such as displaced false teeth, gum, mud or sand, which would prevent air from passing through the mouth.
2. Loosen any clothing that appears to be tight around the chest or around the neck.

3. Place the victim on his back, placing pillows or other support under his shoulders. Tilt the head back so the chin points upward, and pull or push the jaw into a jutting out position.

4. Open your mouth wide and place it tightly over the victim's mouth while pinching the victim's nostrils shut. If the mouth has been injured or if it should be completely clenched, place your mouth over the victim's nose making sure that the mouth is closed. Blow into the victim's mouth or nose.

5. Remove your mouth and listen for the outward rush of air which will indicate that there is an adequate exchange of air. If there is no exchange, gently explore the mouth for any obstruction which you may be able to remove with a gentle sweep of the fingers. Should the victim's tongue be in the way, carefully pull it forward with your fingers.

6. In giving mouth-to-mouth respiration to an adult, you should blow vigorously at about twelve breaths per minute. When the victim is a child, about twenty shallow breaths should be taken per minute.

7. Respiration efforts should be continued until the victim is very evidently breathing on his own or until competent medical personnel have taken over or have directed you to discontinue your efforts. In the event you have discontinued giving artificial respiration due to the victim breathing on his own, you should closely watch the victim to insure there is no further stoppage of breath until medical personnel have taken over.

The foregoing method of giving artificial respiration, although considered the best, is not the only method available. Any method which can be effectively used in an emergency situation is better than no action at all.

HEART STOPPAGE

There is a treatment for insuring that blood from the heart continues to flow, thus sustaining the life process, in the event of a heart stoppage. This procedure is known as Cardiopulmonary Resuscitation (CPR). CPR should be used only by trained and qualified personnel and should never be attempted by anyone who has not received formal training in its application. To attempt this procedure without the proper training could result in additional injury to the victim of heart stoppage.

CPR involves the opening of the victim's airway, restoring his circulation and breathing and the application of a definite therapy. Due to the requirement for formal training before using this procedure, methods which are employed in applying CPR will not be further discussed here.

BLEEDING AND HEMORRHAGE

It is fortunate that most wounds do not bleed profusely. A firm compress applied to the wound and elevation of the wound, where possible, coupled with natural clotting will generally stop the bleeding. However, wounds that are severely bleeding must have first aid measures applied immediately to prevent the loss of life from the bleeding itself or from the resulting shock to the body process.

There are three types of bleeding which need to be recognized in order that the seriousness of the wound can be determined and to dictate the type of treatment that should be applied:

1. Venous bleeding is recognized by its dark red color and the slow steady flow. This type of bleeding is not as dangerous to the life process as the other types listed below. Venous bleeding can generally be controlled by direct pressure applied with a sterile compress or by finger pressure applied to a pressure point between the wound and the heart.
2. Capillary bleeding is typified by blood which oozes from the wound, and usually the application of a sterile compress directly to the wound will stop the bleeding. If the wound is extensive, a tourniquet may be needed.
3. Arterial bleeding is characterized by the blood spurting or gushing from the wound in a bright red stream. This is the most dangerous type of bleeding since the heart is pumping the blood out of the severed artery. In order to stop arterial bleeding, when other methods have failed, pressure must be applied to a pressure point between the wound and the heart, by the fingers or by the application of a tourniquet.

There are certain places on the body where the arteries lie close to the surface and over a bone. These places are called pressure points. By applying pressure to these points you can successfully stop the flow of blood beyond the points used. When conducting first aid classes for security force personnel, be sure that they are aware of where each pressure point is located and what part of the body would be deprived of blood when pressure is applied at any given point. It should be stressed that neither pressure point nor tourniquet use should be applied except as a last resort to stop serious bleeding.

Pressure points are located as follows. For an open arm wound, on the inside of the upper arm in the groove between the large muscle masses—using the flat part of the finger, exert pressure on the artery against the arm bone. For an open leg wound, on the thigh where the artery crosses over the pelvic bone—use the heel of the hand over the pressure point and lean forward over the arm to apply sufficient pressure to stop the bleeding.

Tourniquets should be used only when serious bleeding is taking place and there is no other method of controlling the bleeding. Once applied, the tourniquet should be left in place until removed by competent medical authority. Whenever a tourniquet is used, the location of the tourniquet and the time it was put on should be noted in an obvious place where it is sure to be seen by medical personnel.

POISONING

Substances which are poison to the human system come in many forms. These range from food poisoning to radiation poisoning. The problem of poisons is generally divided into two categories: external poisons and internal poisons.

External poisons result in damage to the skin and eyes, normally in the form of serious burns. These poisons include most acids, gasses which cause blistering of the skin and burning metals, such as phosphorus. In the industrial setting external poisoning will most likely be encountered from chemcials which are used in the production processes or which have been spilled or released by accidental explosion.

Safety equipment, such as goggles, gloves, and protective clothing, should be required whenever working around dangerous chemicals, acids or burning metals. Patrolling security officers should be well versed on the type and extent of safety equipment that is required in various departments of the facility and should report any violation of rules requiring the use of safety clothing or equipment. The best first aid that can be rendered by security personnel is that of prevention of accidents or serious injury to employees by enforcing the rules laid down in this section of the company work policies.

To treat external poisoning, the poison should be diluted by rinsing the affected area with large quantities of water. If the poisoning agent was known to be carbolic acid, the rinse should be accomplished with alcohol, if it is immediately available. A mild ointment and sterile bandage may be applied. A doctor should be obtained as soon as possible. If the eyes have been affected, after the poison has been diluted a drop or two of olive oil or castor oil may be used.

Internal poisoning results from the ingestion of acids, gasses, drugs, spoiled food or any other substance which causes tissue burns or the reduction or slowing of the body functions. Security officers will more likely be concerned with victims that have been poisoned as a result of exaust fumes, chemical or paint fumes and accidentally ingested chemicals rather than victims of drug overdose or food poisoning.

Workers who are required to work in the area of dangerous fumes should be provided with and required to wear masks that are designed to filter out the dangerous gasses or fumes. Workers who have been overcome by such fumes must be moved out of the area into the fresh air and, if necessary, artificial respiration should be administered.

To treat for an ingested non-corrosive poison, the mouth and facial areas should be thoroughly flushed with water. In the event the poison ingested was a corrosive acid, which has burned the mouth, flushing is of little use since the damage has already been done. If the poison has been identified and the proper antidote is available, it should be administered. If the poison has not been identified, the substance should be taken to the medical facility, with the victim, for analysis.

The victim of an ingested non-corrosive poison should be given plenty of water to drink and should be induced to vomit in order that the stomach might be rid of the poison. This can be accomplished by having the victim stick his finger down his throat, having him drink a quantity of warm soapy water, or, if that is not available, he should be given a large quantity of warm water. In this manner, if the poison is not brought up, it will at least be greatly diluted.

The victim of a corrosive poison should be transported to where medical assistance can be obtained as soon as possible.

SHOCK

Any serious injury produces shock to the body functions which is brought on by either mental or physical injury or, in many cases, by a combination of the two. Once a person enters a state of shock, unless it is treated he will go into deeper and deeper states of the shock condition which can eventually lead to death.

A state of shock slows down the circulation of the blood, lowers the blood pressure and results in insufficient blood reaching the brain, other vital organs and the extremities. If the injury has resulted in a major loss of blood, the condition is even more aggravated.

This loss of normal circulation produces the following symptoms which will alert you that the victim is going into a state of shock:

1. Weak and rapid pulse.
2. Skin that is cool to the touch with a clammy perspiration that is particularly noticeable on the forehead.
3. The face, lips and fingernails will pale and possibly turn blue.
4. The victim will chill even in a warm area or climate.
5. The mental faculties will dull, the eyes will droop and half close, the victim will be extremely weary.
6. In some cases, the victim will become nauseous and his breathing may become irregular.
7. If there has been extensive bleeding the victim may be unusually thirsty.

The treatment for a victim of shock is as simple as it is important. The injured person should lie down with the head lowered or the legs elevated to assist the flow of blood to the brain. Tight clothing should be loosened and the victim should be wrapped in blankets or clothing to maintain body heat. Excessive heat should be avoided as it could be dangerous to the victim.

A victim should not be permitted to see his injuries, particularly any extensive bleeding. He should be advised that he will be all right, that he is receiving first aid and that medical attention will be forthcoming. Never discuss or allow a discussion of the victim's injuries within the hearing of the victim. Telling a seriously injured person the extent of his injuries can hasten his death by sending him into deep shock, just as surely as failing to stop his bleeding, give artificial respiration or take other life saving measures.

Members of a security force may never be called upon to render first aid to a seriously injured person but in the event that they are, the victim, their employer and the workers they are hired to protect will expect and demand that they are well versed in the administration of first aid.

QUESTIONS FOR DISCUSSION

1. Define first aid.
2. Why are security officers normally expected to know first aid?
3. What are the basic life and death emergencies which a security officer may be called upon to handle?
4. What action can be taken to restore breathing following a stoppage?
5. What is considered the best method of administering artificial respiration?
6. What is the first action that should be taken upon discovery of a medical emergency?
7. Can CPR techniques be safely used by an untrained person?
8. What are the types of bleeding and how are they recognized?
9. Are different treatments required for the ingestion of a corrosive and a non-corrosive poison?
10. Should treatment for shock be rendered to all seriously injured persons?

MAJOR CONSIDERATIONS

1. For the security officer, possessing the knowledge and ability to provide first aid to the injured is essential.
2. CPR techniques should never be attempted by the untrained.

Reading a book or seeing the technique applied on television is not sufficient training.

3.   The state of shock, in serious injury cases, contributes more to the cause of death than any other factor.

Chapter 20

BOMB THREATS AND CIVIL DISTURBANCES

Although bomb threats and civil disturbances should be planned for as part of the overall emergency procedure plan, their potential for personal injury and property damage is so great that they need to be given special consideration. It should be pointed out that in preparing plans and training personnel for the various contingencies, it is necessary to plan and train for the worst possible catastrophe. A security force that is prepared for the worst will encounter no difficulty in handling the more minor situations which they will be called upon to handle during the course of their day-to-day duties.

BOMB THREATS

The mere threat of a bomb having been planted in a public place may cause panic and injury if the procedures for handling such threats are not well organized and disseminated. Telephonic threats that a bomb has been planted have become far more frequent than the actual planting and explosion of such instruments. It is certainly fortunate that this is true, but this fact makes a plan of action to deal with such threats much more difficult to devise and implement.

The vast majority of bomb threats are received over the telephone, but they may also be received in written form. In most cases where a note is left concerning the bomb, it is delivered to various segments of the news media rather than to the target building or facility. The threat in cases such as this is usually not given as a warning of an impending explosion, but instead contains the reasons for the bombing and identifies the group which wants to "take credit" for the bombing. The delivery of the note in these cases usually coincides with the time of the explosion or is delivered immediately following the explosion. Although these notes may be used as leads or evidence in identifying and ultimately convicting the parties responsible for the bombing, they have no value in the prevention of the particular bombing to which they refer. Therefore, this chapter will concentrate on those threats which are received in time to take positive action to locate and neutralize the explosive, evacuate the affected area, or determine that the threat is only a hoax.

Handling the Call or Note

Since a plan of action is initiated upon receipt of information that an explosive device has been planted in a facility, initial plans and procedures must focus on the people most likely to receive the threat. The caller will not want to be identified and will normally not allow sufficient time for his call to be traced. Therefore, most bomb threat calls will be short and to the point and will be given to the first person to answer the phone. It is recommended that all employees who have occasion to answer any outside phone lines be trained in the proper procedures for recording such calls and in the action that is to be taken upon their receipt. Extensive training should be given to PBX operators and, particularly in those facilities where calls are directed to the security center after normal business hours, to the security guard force.

The training should stress the calm, intelligent handling of the caller, with the goal of obtaining as much information as possible concerning the explosive device. An attempt should be made to obtain information as to the location, size or explosive force and the time the device is set to detonate. For this purpose, a simple, easy to follow checklist should be prepared and should be available at every phone in the facility which might receive an outside call. This information is essential and although it is recommended that the recipient of the call also record as much information as possible about the caller and his background for later identification, obtaining information concerning the explosive device should always take priority.

It is recommended that a recording device be provided which could be activated upon receipt of a bomb threat. The recording can later be reviewed for any background noises, voice inflections, speech impediments, accents or any other information which might help identify the caller or the location from which the call was made. With such a device in place and activated, the person receiving the call can give his total concentration to obtaining all possible information on the explosive device.

Some of the questions which should be included in a bomb threat checklist would be as follows:

1.   What is the specific location of the bomb?
2.   What time is it set to explode?

3. What type of bomb is it? How powerful is the bomb?
4. How will it be detonated? If by timer, is it set for the local time zone?
5. Who are you? What is your connection with the bomb? How do you know about the bomb?
6. Who should I say planted the bomb? For what purpose has the bomb been planted?

It may seem unlikely that the caller will answer such questions, but since the main purpose of such calls is usually to alert any occupants and decrease the possibility of injury or death, the caller may give the essential information, if asked. Most bombings are directed against the establishment and are more often intended to do property damage than to kill or injure personnel. If the intent is to accomplish both, the chances of receiving a call prior to the explosion are slight if not non-existent.

In the absence of a recording device and in addition to background noises, voice inflections and so on, the bomb threat checklist should include questions to be filled in by the recipient of the call providing as much identifying data as possible, such as:

1. Based on the voice, was the caller male or female?
2. Was the caller young or old?
3. Did the caller seem agitated? Calm? Rational?
4. On what phone line was the call received?
5. What were the exact words used by the caller?
6. If information on the explosive device was obtained, did the caller seem familiar with the terms detonator, fuse, timer or charge?

In addition, the check list should provide a place for the full identification of the person who received the call.

Evacuation

As with all other sections of an emergency plan, the person in overall charge of the facility will need to delegate authority and responsibility to act in his absence if a bomb threat is received. He will have to designate what members of the staff will be involved in the resolution of bomb threats, how they are to be involved, who will have the authority to order evacuations, and on what factors the decision to evacuate should be based. If established emergency teams are to be used for handling bomb threat incidents, then that part of the planning will be completed.

The first priority will be to decide whether or not to evacuate. This decision must be reserved for members of top management unless they are impossible to contact, in the time available, before the possible detonation of the device. Failing to evacuate may cost countless lives. However, moving personnel from their workplace may actually place them in greater danger if the location of the alleged device is not known. Evacuation each time a threat is received will be costly in lost production and/or in that the affected employees, guests or customers will be annoyed and concerned about the frequency of the threats. In arriving at this decision the following factors need to be considered:

1. Is the time of detonation known? If known, is there time to safely evacuate personnel?
2. Have similar or identical calls been received in the past and found to be false alarms? Almost 98% of all bomb threats are false alarms.
3. Is there sufficient time to conduct a search of the facility prior to making a decision to evacuate?
4. What has been the experience of this facility as to the frequency of threats and the actual discovery of an explosive device?
5. Do you know the area in which the bomb has been said to have been placed? If known, can that area be safely evacuated without requiring the evacuation of the entire facility?
6. Do you have any disgruntled former employees who might be responsible? Would they be likely to have had the opportunity to plant an explosive device? Would they be the type who would carry out such a threat?
7. How good is your protective security system? Could an outsider penetrate the security screen and plant an explosive device? If so, what areas within the facility would be the easiest to penetrate? In those areas, where would the device most likely go undetected?
8. Did the caller reporting the bomb seem nervous, agitated, angry or otherwise upset?
9. Based on past experience, is it your judgment that the caller is perpetrating a hoax?

When a decision is made that evacuation is necessary, the same procedures used for other emergency situations will be followed. If existing procedures have been tested and all personnel are thoroughly versed in their routes and have confidence in the evacuation plan, safe evacuation of the facility without causing a panic among the personnel should be possible.

Understanding what motivates people to call in a false bomb threat might be of assistance in arriving at a decision concerning the advisability of evacuation. The reasons are many and some would even be humorous if they did not have such a great capacity for injury and lost earnings.

Take for example the manufacturing plant in central Illinois which received a bomb threat from a male caller. It was later learned that the caller had been given the day off from his own job and wanted his lady friend, who worked in the plant, to be able to join him for a day of leisure. Schools often receive calls made by or instigated by students who hope that classes will be cancelled, thus giving them a free day.

Most calls are made for more basic reasons such as revenge for an actual or supposed wrong, harassment of a business competitor, disruption of production out of jealousy, or a prejudice aimed at the management of a facility.

Outside Assistance

The search for the alleged device must be extremely thorough and conducted in the shortest possible time. Although there are several methods which can be used to select personnel who are to be assigned to conduct the search, the best method and the only one recommended here is that the search teams be selected on the basis of their knowledge of the area to which they are assigned. Once selected, they must be given a complete training program on what to look for and what action they should take should they encounter a suspected explosive device.

Following the emergency team plan, the risk manager would coordinate all requests for outside assistance and would be responsible for obtaining standby medical personnel or equipment and for notifying bomb disposal units. The plan should include the location and telephone numbers of the nearest available unit. Most city police departments, or fire departments, now include bomb disposal units. Assistance is also available from explosive ordnance demolition units stationed at most military bases. The risk manager would also be responsible for the coordination and operation of the training program to acquaint selected search team personnel with the methods of search and the do's and don'ts of handling suspected explosive devices. Police, fire or military bomb squad organizations are normally available to conduct this training and have examples of various types of devices for display.

Contingency plans need to be included in the actions to be taken in the event personnel are trapped in an area as the result of an explosion. This should include the quickest method of gaining access to any area within the facility in the event the normal entry ways and exits have been blocked.

No attempt should be made to move a suspected device or to detonate a known device except by qualified, expert members of a bomb disposal unit. Many devices are designed to explode at the slightest movement and inexpert detonation will likely cause extensive damage and personal injury.

The author has had some experience with bombs and bomb threats and would like to pass on several of these incidents to make a point. While in the former city of Saigon, South Vietnam, we were reading in our living quarters when a knock on the door interrupted us. The door opened, revealing one of the military police guards who calmly and politely requested that we leave the building for a short time. He went on to tell us that a bomb had been found on the floor below and the demolition team had wanted the building evacuated before trying to deactivate or remove the device. The deliberate and calm manner in which this M.P. handled the evacuation of the building had a calming effect on the residents and made the evacuation a safe and orderly procedure.

The second incident took place in a busy, crowded casino of a major resort hotel in Las Vegas. The security force was advised that a bomb threat had been received alleging that a bomb had been planted in the casino and was set to detonate in 5 minutes. There was no time to attempt an evacuation of the casino, but a preplanned search of the casino area was started using security and management personnel designated in the emergency procedures plan. Fortunately the call was a hoax. Due to the existence of emergency plans and procedures to deal with such situations, no evacuation was attempted and the possibility of panic-produced injuries was avoided. If you have ever been to Las Vegas and noticed that most members of the casino security forces are above average in size, it could be because that in any workable evacuation plan, the crap shooter or blackjack player who is on a hot streak would have to be physically removed from the table. There is no way they would leave voluntarily even if told that an atomic device had been planted under their table.

CIVIL DISTURBANCES

Unless caused by a labor dispute, the average industrial complex, office building, or wholesale or retail business will not be the target for mob violence during civil disturbances. This is not to say that they do not need a plan of action to provide additional protection during riotous situations. It is often the innocent bystander or business that receives the most serious damage during such disturbances.

A riot or mob violence, to a lesser degree, is usually brought about through dissension with a government policy, problems between races, religious groups, political groups or dissatisfaction with the establishment. The government at which the violence is directed can be that of a university, city, county or any other political unit. Consequently, mob violence has been directed at banks, schools, atomic-powered electrical generating plants, military installations, manufacturing plants, offices of multi-national corporations, chemical plants and government offices at all levels.

Part of any emergency procedures plan will have to provide for counteractions designed to protect the lives and property at the facility. These actions need to be prepared for the possibility that the facility might become the target of rioters as well as for the protection and continued operation of the facility during such occurrences, even if the facility itself is not a target for attack.

Intelligence Gathering

When unrest is evident in the surrounding community and the possibility exists that it might grow into violent confrontations between opposing groups, intelligence-gathering procedures need to be implemented to keep management abreast of the situation as it develops. During this period close liaison is required between the facility and area law enforcement, fire fighting and medical units which might be called upon to provide assistance. The risk manager or security department head, as before, will be responsible for liaison with these agencies and will most likely be designated the responsibility of providing the intelligence-gathering effort.

Role of the Security Department

The plan should include provisions for calling in all off-duty security personnel and for supplementing the ranks of the security department with other members of the emergency response team. The special training given to all personnel assigned duties during civil disturbances should detail actions that should be taken during a confrontation and the actions or type of actions which are to be avoided. These actions and restrictions should reflect company policy within the framework of the legal remedies available to the firm. In any event, mob control and dispersal should be left to the police or military riot control unit which has been given the responsibility of quelling the violence. Should it be necessary to take action against a mob to prevent injury, pending the arrival of the authorities, then the action should be restricted to only the degree necessary for the protection of personnel. Damaged property can be repaired or replaced but human life is not as yet within our power to replace.

Riot Control

A mob is formed when a group of people who have gathered in a crowd for a common purpose, usually a form of dissent but not always, lose their individual identities and become members of the mass following the lead of one or more activists. Members of a mob feel secure in their anonymity and justify their violent acts as being acts of the mass rather than their own individual acts. Mob control formations and procedures are designed to strip away the mask of anonymity and return the members of the mob to rational thinking individuals. The theory of divide and conquer is used in taking action to disperse the mob through increasingly smaller division.

The actions that are taken by a riot control unit in dispersing a mob will be listed. It should be made clear that these are options used by a police or military unit charged with quelling the riot and restoring peace to the community. They are not recommended for inclusion in the emergency procedure plans since, as stated, the control and dispersion of a mob is the job of the assigned law enforcement unit. The riot control unit is required to accomplish the dispersal of the mob using only that degree of force which is necessary. To accomplish this they set into motion a series of pre-planned actions known as the priorities of force. The action on the part of the riot control force stops at whatever point their objective has been achieved. The priority of force usually consists of the following and the degree of implementation will be decided by the commander, based on the threat to the well being of the community posed by the riotous group.

Show of Force.   The riot control unit is formed and brought into view of the rioters.

Issuance of a Proclamation.   The crowd is read a proclamation, usually ordered by the governor of the state, stating that they are part of an illegal gathering and they are ordered to return to their homes.

Use of Riot Control Formations.   By entering the crowd using a wedge or other similar formation, the riot control force attempts to split the mob into smaller and smaller segments until they have been effectively dispersed.

Use of Chemicals or High-Pressure Water Hoses.   By using tear gas or vomiting gas or applying a spray from a high-pressure water hose, an attempt is made to split up and disperse the gathering. In this phase, as in the one above, it is necessary that an avenue of escape exists for members of the crowd.

Identifying Leaders and Taking Them Into Custody.   After identifying leaders of the mob, the mob is infiltrated to take the leaders into custody or a selected formation is used to reach the leaders and take them into custody.

Use of Selected Marksmen.   After identifying the leaders of the mob and failing in the attempt to take them into custody, selected marksmen are used to bring fire to bear on the leaders.

Use of Full Fire Power. This is used only as a last resort when all other efforts have failed and the mob is engaged in life threatening activities. The full fire power of the riot control team is brought to bear on the body of the mob.

It is very unlikely that members of the security force or facility emergency team would ever be called upon to break up a riot. However, some of the techniques used may be helpful in containing a mob until the arrival of the authorities. The first action that could be taken by facility management is to call upon the leaders of the mob to enter the protected area for a discussion of their grievances. Failing in this, they might address the gathering, inform them that the authorities have been called and are en route and ask them to disperse and return to their homes.

The facility fire control team could be brought up with their equipment and, if required, could direct streams of high-pressure water toward locations where members of the mob were attempting to enter the property. Plans should include provisions for pulling the security force and emergency team members back into the protection of the buildings if efforts should fail to prevent the rioters from entering the property prior to the arrival of police.

Planning should also include provisions for special equipment which might be needed for the protection of personnel during a civil disturbance. Such items as hard hats, flack vests, gas masks, portable barriers, portable lighting or other similar protective equipment.

The business or facility which is not prepared to cope with bomb threats or civil disturbances is taking the unnecessary risk of exposing property and personnel to damage or injury. Being prepared to handle such contingencies may not eliminate the possibility of their occurrence but it goes a long way in minimizing loss of life and property.

QUESTIONS FOR DISCUSSION

1. What is the purpose of obtaining a tape recording of telephonic bomb threats?
2. If a search of the premises fails to disclose an explosive device, can the threat then be ignored?
3. What are some of the questions which should be asked of a caller reporting that an explosive device has been planted in your facility?
4. Discuss the factors to be considered when making a decision concerning the evacuation of the facility or the designated area where a bomb has allegedly been placed?
5. Why is it important that all employees be familiar with evacuation plans and procedures to be implemented in the event of receipt of a bomb threat?

6. What is the purpose of intelligence gathering during periods of civil unrest?
7. What official will be responsible for liaison with outside agencies and gathering intelligence information when civil disturbances are expected?
8. Why should it be necessary to establish restrictions on actions that can be taken during a confrontation with a mob?
9. Discuss the types of facilities which might become targets of mob action during periods of civil unrest.
10. What are the factors that permit a member of a mob to engage in violent, unlawful acts which he would not dare if he were acting alone?

MAJOR CONSIDERATIONS

1. When plans are prepared and personnel trained for the worst possible situation, it usually follows that the more minor problems which are encountered are able to be handled with comparative ease.
2. Although the vast majority of bomb threats turn out to be false alarms, we cannot afford to ignore a threat and must analyze and decide our course of action on each occurrence independently.
3. Civil disturbances or riots may never be a problem for a given facility, just as lightning may never strike the same place twice, but it is better to be prepared for the eventuality of the happening rather than suffer the consequences.

Chapter 21

EXECUTIVE PROTECTION

There was a time, only a few years ago, when the term executive protection applied only to the protection afforded to heads of state, their cabinets, ministers and other high-placed government officials. The idea that business executives might need extra protection from kidnap/extortion or assassination attempts began when such attacks became the tool of political activist groups such as the Symbionese Liberation Army, Palestine Liberation Organization or the Red Brigade.

TERRORISM

Terrorism is defined as the use of terror, violence and intimidation to achieve an end. Terrorism takes many forms. Adolph Hitler effectively used terror to intimidate the German people and keep them under control during his reign in Nazi Germany. True, his prime target was the Jewish people of Europe, but anyone who dared disagree with Hitler was guaranteed, at least, a trip to the infamous concentration camp. Dictators, throughout the history of civilization, have used this tactic to achieve their goal of mastery over the populace that they rule. Italy's Red Brigade is a small group of persons bent on imposing their will and their beliefs on the entire country by attempting to force the political leaders to yield to their demands that they deserve a major voice in government despite the fact that they do not represent enough of the populace to permit them to be elected as representatives of the people.

The holding of 53 U.S. citizens, as hostages, by the country of Iran is another classic example of terrorism. It is not the kidnapping or holding of someone that really constitutes a terrorist act. The threat that the person(s) held hostage will be physically mistreated or killed and the uncertainty concerning their well being and ultimate fate, which makes us want to take whatever action that is necessary, even submission to unreasonable demands, makes the act effective.

The demands which are made as conditions for the release of hostage(s) vary according to the needs and desires of the group responsible for the kidnapping. Generally they are two-fold: money and the release of "political prisoners". Businessmen representing major multi-national corporations are prime targets because the firms they represent can provide the money requested and can also exert a certain amount of pressure on the government holding the alleged political prisoners. It is not our purpose to judge whether a cause is just or unjust, for in the taking of hostages the group contaminates their cause by their use of a tactic that is patently unlawful and morally indefensible. Just as protection of assets is accomplished through loss prevention, protection of executive personnel must be accomplished by establishing preventative measures designed to make it as difficult as possible for the terrorist to achieve his goals. To help explain where security personnel fit into the overall program for executive protection, the establishment and operation of an effective program will be outlined below.

ESTABLISHING A GUARD FORCE

The first decision that must be made is much the same as selecting a method of establishing a security guard force. That is, do you hire your own experts and set up a proprietary program, do you contract with a company that provides such expertise or will you contract with consultants to provide the expertise and then utilize proprietary personnel to operate the program? Under the principles of risk management, if an effective program can be obtained that costs less than another option, then that would be the program to adopt. The key word is effective. Any program that is not effective can be considered a loss, regardless of its bargain basement price.

ACCESS TO INFORMATION

In any case, one of the first considerations in setting up or improving an executive protection program would be deciding which personnel will be given authorized access to the information gathered and the organizational structure of the program. This information should be disseminated on a strict *need to know* basis so that only those working on the program have access to the information that is necessary for them to properly perform their duties.

All personnel who are to be authorized access to executive protection plans or information should be the subject of a thorough background investigation which should be updated at, at least, six-month intervals. Background investigations should include, in addition to financial history and past history of the person being investigated, his social activities, relatives, contacts and habits which might make him susceptible to blackmail or the acceptance of a bribe.

Employees who are being considered for positions which will require access to this information should be advised of the requirement for the background investigation as well as the extent of such investigations. They can then be given the opportunity to refuse the position or to voluntarily agree to the investigation. This should eliminate any problem which might arise out of alleged invasion of a person's privacy.

EXECUTIVE PROFILES

The next step in the program is to identify the executives or executive positions for which protection will be extended and to develop personal profiles on all of the executives included. These profiles will include each member of the executive's immediate family, and all information collected concerning the executive will be duplicated for each family member.

Each profile will consist of the following:

1. Full name to include nicknames or family pet names.
2. Address of primary place of residence and any other places or residence owned or used by the executive or members of his family.
3. Full physical description.
4. Date and place of birth.
5. Recent color photographs: full length, frontal and profile close-ups, all of which should be kept current.
6. Outside interests, sports or hobbies.
7. Medical history including any required medication or recurring treatments.
8. Daily routine including names and addresses of schools, summer camps, modes of transportation and any other pertinent information.
9. Full descriptions of motor vehicles owned or used including license numbers, state in which licensed and vehicle identification number.
10. List, fully identify and give relationship of all close personal friends or associates including their residence addresses, business interests, political leanings and social activities.
11. List an occurance in the life of each person which would not generally be known outside the family circle, to be used to positively identify the person and insure that he is well and functioning.

12. Code names and phrases should be established, for each person, and made a part of his profile. These can be used to identify location or direction or merely to indicate that the person is well and not injured.

From the information contained in these profiles, it is not difficult to understand why it is necessary that they be given the highest possible security protection. The criteria established for background investigations of personnel having access to these files may seem less stringent when the nature of the information that will be handled is considered.

PROTECTING THE RESIDENCE

While compiling profiles, security surveys can be instituted on all residences that will be provided protection to bring them up to a *site hardened* condition. In providing protection to a home, consideration must be given to the same factors that would be considered in the protection of a business site, but the barriers and other security hardware would probably differ due to the desire to avoid the appearance of an armed camp or prison.

Perimeter protection, for instance, may be provided by a more aesthetically pleasing brick or concrete block wall of eight feet or more, which will also provide a greater degree of privacy. In place of the barbed wire overhang, the wall can be topped with spikes or cut glass imbedded in cement. In some cases the perimeter will consist of the exterior walls of the building, as would be the case if the residence were located in a townhouse, condominium or apartment building.

An identification system would have to be established for members of the household staff, groundskeepers and security personnel. An extension of the business' decal system could be used to identify authorized vehicles. Up-to-date guest lists would be required for recognizing and identifying bona fide visitors. Gates allowing entry to the property would most likely have a sally port included, be electronically operated and be conrolled by a security officer who would demand positive identification prior to opening the inner gate.

An effective protective lighting system would have to be installed, as would alarm systems. This might include pressure-type alarms inside the perimeter barrier and near windows and doors of the residence, as well as magnetic alarm devices installed on the doors and windows proper. The installation of closed circuit television to cover the grounds, gates, perimeter and selected areas within the residence would go a long way in improving the security of the property. Of course, dead bolt locking devices imbedded in reinforced frames and sills on all doors and windows of the residence would have to be included.

Guard force response to alarms would need to be provided with sufficient manpower to deter an asssault on the property or to delay an assault until the police had time to respond. Security patrols would probably be conducted with guard dogs accompanying the patrolling officers.

Motor Vehicles

A protected area should be provided where the executive and his family could enter and exit motor vehicles without being exposed to possible sniper fire. Vehicle maintenance should be provided on the premises or the business site whenever possible. Vehicles serviced in other areas would require a member of the security force to be with them at all times to preclude the possibility of an explosive device being planted in or on the vehicle. All motor vehicles used by the executives and their protective force should be armored to prevent small arms fire from injuring the occupants. They should also have firing ports provided so that a protective fire could be directed upon potential attackers.

Training

Training classes should be established for executives, chauffeurs, bodyguards, escort guards, residential security officers and adult members of the executive's family as well as household staff members. Executives, chauffeurs and all security personnel should be trained in the techniques of evasive driving using the armored vehicles, which are much more difficult to handle than conventional motor vehicles, and in the effective use of the various types of firearms selected for inclusion in the protective program. Adult family members and other staff members could be optionally trained in these subjects. All security personnel, adult family members and other staff personnel should also be trained in para-medical techniques and procedures.

The evasive driving course should include instruction on the use and placement of escort vehicles leading and following the protected vehicle, the use of decoy vehicles and the problems encountered if a discernible pattern is established by using the same routes to and from the residence. Practical exercises need to be held in using evasive techniques, ramming, preplanned escape routes, high-speed turns and other elements that may be needed for the protection of the executive while in his motor vehicle.

As a case in point, former Italian Premier Aldo Moro was a creature of habit whose punctuality and predictability made it easy for his abductors. His driver followed the same route each day, allowing the terrorist to lay in wait at an intersection he knew the premier would be passing. Members of the Red Brigade's attack force had been well trained and their attack well planned and executed. They had slashed the tires of an automobile owned by a vendor who normally did business at the selected intersection to insure he was not there on the day of the attack. All telephone lines in the vicinity of the selected point of attack had been cut to prevent an alarm from being given. There was no lead vehicle escorting Mr. Moro, and neither the vehicle in which he rode nor his escort vehicle had been armored. The selected point of attack was a regular bus stop and the attackers were dressed in the uniform of Italy's domestic airline. Even if the drivers has been trained in evasive driving techniques, they probably would not have had time to use this knowledge. There had not even been time for a radio alarm to be given, since the five people with Mr. Moro were dead before they knew what had taken place. The abductors took the wounded Moro to a waiting escape vehicle and fled the area. An effective executive protection plan coupled with the proper equipment and training could have written a different end to this story.

PROTECTING THE WORKPLACE

As with the residence, a secure place must be provided, at the workplace, for the executive to enter or exit the motor vehicle. This can be accomplished by using physical barriers or by insuring that the executive is surrounded by protective personnel any time he might be exposed to potential hostile fire.

Entry control methods adopted for the protection of executive offices, conference rooms and dining areas must be foolproof. This requires positive identification and authorization of anyone who is to be permitted to enter these areas. To accomplish this, without offending business visitors, dignitaries or friends, will require a delicate balance of strict security procedures and tact.

As previously stated, the prime targets for today's political activists are the representatives of the multi-national corporations. For the most part, kidnappings have been conducted at offices and facilities in countries other than where the corporation is headquartered. Although there is no guarantee that this will continue to be the case, executive travel or residency in a foreign country must receive special attention if adequate protection is to be provided.

TRAVEL

Travel itineraries must receive the highest security classification, and any prior publication of travel plans should be cleared with the risk manager. It would be helpful if codes were adopted in the advance planning for executive travel. Site hardening of out of country business sites and residences must be particularly well planned and executed. Local travel within the country must be unpredictable and well protected, with times and destinations highly classified on a strict need-to-know basis only.

Intelligence gathering concerning areas in which executive travel is to take place or where business operations are conducted must be thorough and constantly updated. Liaison with government departments, local police officials and with INTERPOL should provide information on terrorist groups and their activities which will permit the necessary degree of travel or site security to be implemented.

KIDNAPPING

In addition to a desire for money or the attainment of a political goal, most activist groups have another driving objective—the need to publicize their cause and the goals they want to achieve. This objective will, at times, overrule all other considerations. Due to this need for publicity, the security planner may never rule out the possibility of the executive being the target of an assassination rather than kidnapping.

Executives who are included in the protective program should be advised of the actions they should take in the event the protective screen was breached and they were in fact kidnapped. They should be told not to offer any resistance or attempt to escape once they are in the custody of the kidnappers, unless it is evident that their abductors do not intend to free them but will kill them even if their demands have been met.

This determination should not be made on conjecture since the kidnap victim's best chance of survival is in his cooperation with his abductors. In the normal, profit-motivated criminal kidnapping, a good indication would be that the kidnappers never allowed the victim to see their faces, which would indicate that they probably intended to release him once their demands have been met. In terrorist group kidnappings, this may not be a valid indicator since they will normally identify their group and most members of the group will already be known to law enforcement authorities. In this case, it would not matter that the victim could later identify his abductors.

If, as in many cases, the ransom demands not only include money but also the demand for the release of political prisoners or members of the terrorist group, and the government involved refuses to release such prisoners, the victim should do whatever he can to escape or obtain his release. In most hostage situations it has been found that a psychological bond is formed between the abductors and their captive. The victim can use this bond to his advantage by attempting to convince his abductors to settle for the money demanded and the resultant publicity for their cause.

Executive protection planning should include the designation of certain corporate executives and alternates who are to be authorized to negotiate with the abductors for the release of the victim. Risk managers of multi-national or other large corporations whose executives may be targets for kidnap/extortion attempts should be certain that adequate insurance protection is provided for such eventualities.

As can be seen, security professionals are involved in executive protection as consultants, planners, managers, bodyguards, escort officers, and security personnel assigned to protect both the residence and business locations. It is a difficult assignment as is evidenced by the many maimings, kidnappings and assassinations that have been successful. By learning from the mistakes of others and by providing the most effective protection possible, this trend can be reversed and the protection afforded executives can be as viable as the protection provided to other valuable business assets.

QUESTIONS FOR DISCUSSION

1. What factors have made it necessary to provide protection to business executives?
2. What is the first step in starting an effective executive protection program?
3. What security precautions should be taken to safeguard information on the protected executives and their family members?
4. Discuss the information which should be gathered as profile information on the executives and their family members.
5. What steps can be taken to insure the safety of the executive while at home?
6. Would the same type of perimeter barrier that is used for an industrial site be desirable for use at a residence?
7. What action can be taken to provide additional safety to the executive while en route between home and office?
8. What type of training should be conducted and who should be included?
9   Will it be necessary to adopt more stringent security measures in the executive office area since it is already within the protected area?
10. Where and how can intelligence be obtained pertaining to areas outside your native country where executive travel or business sites might be required?

MAJOR CONSIDERATIONS

1. Executive protection operational details, information in executive profiles, travel itineraries and knowledge of which executives are included must receive the highest priority of security safeguarding.

2. Terrorist activities are not limited to organized political activists or dictators; the threat could come from any corner, and the security professional must be prepared to meet the threat.
3. A corporation's executives are among its greatest assets. Even if they were not, the corporation that did not protect its executive personnel would find it next to impossible to recruit other executives to fill positions vacated by a kidnapping.

IV. SPECIAL
CONSIDERATIONS

Chapter 22

Patrolling Techniques

Although it may come as a surprise to some security officers, the purpose of conducting security and safety patrols is not to provide guards with physical exercise, keep them from becoming bored or even to keep them awake. Patrols are necessary to insure the integrity of the overall security program. Frequent and total coverage of the protected area is needed to provide the most timely discovery and correction of security, safety and fire hazards. Patrols also allow for the discovery of fire while still in its incipient stage.

Security patrols, equipped with some method of communication, can be dispatched to accidents and fires, to check on alarms or suspicious activity observed on closed circuit television, or sent to the assistance of other security personnel.

HAZARDS

As discussed in Chapter 17, a safety hazard can be categorized as any condition which would make movement through the area dangerous to security personnel or other members of the work force. This category of hazards would also include employees engaging in horseplay, unsafe operation of motor vehicles, careless handling of dangerous chemicals or materials and the unsafe operation of machinery.

A fire hazard would be any act, omission or condition which makes the ignition or spread of fire likely, including such things as oily rags thrown into a corner where they may spontaneously combust; welding equipment used in areas where paint is stored, or near flammable dust, wood chips, gasoline or other volatile substances; smoking in areas designated as "No Smoking"; or frayed electrical wiring installed in the building or on appliances or machinery. In short, anything that would be likely to cause a fire or which, once started, would give it fuel is considered to be a fire hazard. Fire hazards encountered by patrolling security personnel should be corrected on the spot, if possible, or reported to management and the applicable department for correction.

A security hazard would be any act, omission or condition which breaches the security of the protected area and makes the penetration of the area, by a potential intruder, more likely or which provides the pilferer, saboteur or industry spy access to the area or a lane by which he will be able to remove assets from the area. Not properly securing proprietary information, not locking doors and gates, not promptly repairing or replacing critical light standards or fixtures and not properly protecting or securing high value products or materials are examples of security hazards. A list of all the hazards to the security of a protected area, for which a patrolling officer should be alert, would fill a book in itself. Some of the hazards a security officer is likely to encounter, and which he should be alertly checking for while on routine or safety patrols of the protected area, will be listed here. For ease of recognition they will be listed by specific areas.

Perimeter Barrier

1. Routinely check for any breach of the barrier such as holes in or under the barrier.
2. Where a barbed wire overhang is installed, check for any indication that it has been pushed down, cut or broken by possible intruders. Report any areas where the overhang is in need of repair or replacement due to physical damage or exposure to the weather resulting in rust.
3. Check all unattended gates to see if they are properly locked and to determine if there is any indication that the lock has been tampered with or replaced. Thieves have been known to steal locks which have been left open when the gate is being used or to cut the lock from the chain or hasp. They then replace the lock with one of similar manufacture and appearance for which they have the key. The switch would not normally be noticed until the next time the gate was to be opened, which in some cases could be weeks or even months. To prevent this type of incident, patrolling officers should periodically check the lock on gates by using their key to insure that a switch has not taken place. This switch has been made by thieves

who would take the lock from the gate during the day, making the switch with their own lock. During the night they would enter the area, steal the material they were after and on their way out would replace the original lock on the gate. This went on for several weeks before the switch was discovered. This could have been easily prevented if the lock had been locked on the gate while the gate was opened rather than being left hanging on the gate without being locked.

4. Check the clear area on either side of the perimeter barrier to make sure it is not being overgrown and make a report any time the area needs to be cleared again.

5. Check to insure that there is nothing stacked against or left within 15 to 20 feet of either side of the perimeter barrier which would assist a potential intruder in gaining access to or making a quick exit from the protected area.

6. After a heavy rainfall check the soil along the perimeter barrier for any sign of erosion. Make a report of the condition at the first sign that water drainage is undermining the security of the perimeter barrier.

Outside Areas

1. Check outside storage areas or truck parks for shadowy areas which might provide a hiding place for a potential intruder or which could cover the activities of a pilferer attempting to reach the perimeter to throw pilfered items to a confederate outside the barrier.

2. Check storage areas to insure that they remain organized in such a way as to provide an unobstructed view of the perimeter. Check to make sure that covers and tie downs are properly locked or sealed and that they have not been tampered with.

3. Check for any materials or items that are out of place which a pilferer might be gradually moving toward the perimeter barrier in an attempt to get them outside of the protected area.

4. Check the locks or seals on trucks which are partially or fully loaded to insure that they have been properly secured and have not been tampered with. In the case of refrigerated cargo, check to insure that the temperature of the trailer is within acceptable limits.

5. Check all lamps in the protective lighting system and report any malfunctioning or burned out lamps for correction. If the lamp is out, check to see if the bulb has been broken, and if so, alert the security control center of a possible intrusion attempt.

Building Patrols

1. Check all exterior and interior doors and windows, which should be locked, to make sure that they are locked. Do not take for granted that a door or window is locked because it has been checked and found to be locked on a previous tour. A former president of the United States was ultimately forced to resign his office as the result of a security officer's alertness in making sure a door was actually locked rather than taking it for granted.
2. Check the lights over entrances and in aisles or stairways to make sure they are working. If they are out, check to see if they have been turned off, unscrewed or are burned out. Should they have been unscrewed or turned off since the last tour, notify the security control center and institute a search for a potential intruder. Report all bulbs that have burned out so that they may be replaced.
3. Periodically check the functioning of local alarms installed on emergency exit doors to make sure the battery has not gone dead and that the alarm is functioning. In facilities where proprietary alarm systems have been installed, a periodic check of the functioning of the system should be made. This can be simply accomplished by reporting your position to the control center, opening a door or window that is on the alarm system, and making sure that the monitor receives the alarm on the annunciator panel.
4. Check any broken, damaged or torn open cartons. Note their condition for future reference on later tours and make a report of their condition for correction. A half-full carton noticed on one tour which is empty, missing or only a quarter full on the next tour and no one is working in the area, is a positive indication that a pilferer is active in the area. If employees are working in the area, check with the supervisor in charge of the area to determine if the broken carton was moved or emptied by employees and is therefore accounted for.
5. Security cages, proprietary information storage areas, data processing centers, research and development areas, data research and development areas and all other restricted areas should be checked each tour to insure that they have been properly secured and have not been tampered with or entered by unauthorized personnel.

These are but a few of the hazards that an alert security officer should be looking for when conducting a patrol of the protected area. Each business will have certain areas and items that will require close attention by the patrol officer. Each patrol tour should be as thoroughly conducted as possible and patrols should never be turned into a race to see how fast patrol rounds can be made or to allow more time for resting at the control center between tours.

PATROL SUPERVISION

Patrols of the protected property should be supervised in some manner to insure that the desired coverage is being obtained and that patrols are conducted as thoroughly and frequently as is desired. The supervision may consist of anything that tends to verify that patrol routes are being followed and that each area or location that the guard should visit is actually being visited and inspected by the patrolling officer. Some examples of methods which could be used to supervise the patrols would include having the patrolling officer punch a time clock at various employee clocking areas located throughout the area, using a portable watch clock, keying in on an electrical supervision station or using a magnetic key card in a card reader.

Fire insurance carriers, as an added inducement to insured businesses, will often reduce the cost of the insurance premium where it can be documented that patrols are being made on a frequent basis and are conducted throughout the area. It is possible for the insurance carriers to provide this incentive since the early detection and reporting of fire serves to reduce the ultimate loss from fire and water damage.

Although it is required that certain areas be visited during each patrol tour, the routes which are taken to reach these areas should be varied from tour to tour. This is done in order that potential intruders or pilferers will be unable to predict where the patrolling officer will be at any given time.

Of the supervisory methods listed, only two are in widespread use. These methods and their benefits and drawbacks are described below.

1. Electronic supervision is conducted by patrolling officers visiting and keying electronic keyways strategically located along the patrol route. Electronic stations report to a central station alarm center. This system provides an excellent safety factor for the patrolling officer in that a station that is missed, visited out of order or not keyed within a predetermined time period will result in an alarm being given at the central station. Although the system provides a permanent record and insures that patrol routes are followed, its major drawback is that the patrol officer must follow a precise route and be at a specific location at exact times. This allows anyone planning acts of theft, sabotage or espionage to know exactly where the officer will be at any given time.

2. Portable watch clocks are carried by the patrolling officer who visits key stations along the route. Each station has a key which, when inserted into the clock and turned, records on a disc or tape an imprint indicating the station that was visited. The time of the visit is shown by the location of the imprint on the disc or tape. There is no safety feature for the security officer when using this system, but the key stations may be visited in any sequence, allowing the patrolling officer to vary his route on each tour. A

check of the tape will determine whether all keys were visited on any given tour.

For those businesses which have installed magnetic key card access systems with card readers throughout the facility, supervision of the patrol route can be obtained by having the patrolling officer insert his key card into selected readers along the patrol route. The printout of each reader will indicate that the officer's card was used and will show the time of the visit. This system would also allow the patrolling officer the flexibility of visiting the card readers in any order, thus avoiding the possibility of establishing a discernable pattern.

Patrols which are conducted in widespread areas may require the use of a form of transportation such as the one depicted in Figures 22-1 and 22-2. Motorized patrols can be supervised using the same systems described for foot patrols. Patrol vehicles, when used, should be equipped with emergency portable fire fighting equipment suitable for use against the four classes of fires. A portable or attached searchlight or spotlight of sufficient power to illuminate any area the patrol is required to check and a fully stocked first-aid kit should also be standard equipment. If possible, the vehicle should have a two-way radio installed or the patrol officer should be issued a portable two-way radio of the type depicted in Figure 4-2. Figure 22-3 depicts a vehicle-mounted radio which is compatible with hand-held units used by foot patrols and other members of the security force.

Security officers should strive to make each patrol as thorough as possible using their sense of sight, smell and hearing to detect anything unusual or out of the ordinary. In this manner they will be totally aware of everything that is taking place around them. Alertness is essential to insure the thoroughness of the patrol and the safety of the patrolling officer. Foot patrols should avoid walking too close to building walls to preclude the possibility of a thrown or falling object hitting and injuring them. Darkened door-ways or alcoves should be given a thorough inspection using the hand-held or vehicle lights before being entered by the patrolling officer. Speed laws pertain to the security officer the same as they do to other employees. Vehicle patrols are much more effective at a low rate of speed, and the members of the security force should observe speed laws unless responding to an emergency situation that requires disregarding those laws. Even in such situations, unless the vehicle is equipped with emergency lights and warning devices, the speed limits should be obeyed.

Figure 22-1.   Patrol vehicle utilized for parking control. Courtesy of OMC Lincoln, Cushman-Ryan.

Conducting thorough patrols, particularly during off-shift hours or when the security guards are the only ones present in the facility, is one of the more important duties that security officers must perform if the protection program is to be successful.

QUESTIONS FOR DISCUSSION

1.   What is the purpose of conducting security patrols?

Figure 22-2.   Patrol vehicle showing storage compartment for emergency equipment.
Courtesy of OMC Lincoln, Cushman-Ryan.

Figure 22-3.   Vehicle-mounted two-way radio works in conjunction with hand-held
portable models. Courtesy of Wilson Electronics.

2. What are the three major categories of hazards for which patrolling officers should be alert?
3. What security hazards might be expected to be found on or near the perimeter barrier?
4. Should a patrolling officer be suspicious if he finds that a light bulb has been unscrewed? Why?
5. What are some corrections the patrolling officer can make to improve the security of the protected area?
6. What are some methods of patrol supervision?
7. Of the available patrol supervisory methods, which appears to be the most effective?
8. What are some of the problems encountered when using the electronic method of patrol supervision?
9. List the basic equipment that should be carried by a motorized patrol.
10. Are security patrols really necessary? Why?

MAJOR CONSIDERATIONS

1. Security patrols are the audits and inspections which insure that the other elements of the system are functioning effectively.
2. Patrols should be supervised using a method that will allow patrol routes to be varied, while providing total coverage of the protected property.
3. Security patrols provide for the early detection and correction of hazards and minimize losses from accident, fire or theft.

Chapter 23

Report Writing

It is presumed that ancient cavemen communicated by sounds similar to communications used by animals. Evidence has been found of symbols used in written communications such as a picture of the sky, an animal, a tree or a man, all of which were used to communicate an idea.

Fortunately, for those of us living now, oral and written forms of communication have grown and developed to the point where the words written here can be read and understood by any literate English-speaking person. It can be carried a step further by translating these words into other languages so that they may be read and understood by any literate person on the face of the earth.

The spoken word is a powerful form of communication between people. Unfortunately, it is limited by the fact that the words being spoken must be heard in order that they be understood. It is also a quirk of human nature that a story once told gains or loses something each time it is retold. Today, if we are not glued to the television set, we are listening to the radio or the hi-fi. It is not surprising that we tend to lose sight of the importance of the written word.

Were it not for the written word, there would be no television, no movies and very little in the form of radio broadcasts. The programming for all of these media starts with the written word. Scripts, stage directions, cue cards, camera positions, commercial breaks and even music lyrics and notes all must take the written form.

Nowhere is the written word more important or critical than in the writing of a report. Reporting the facts, as they happened, of an accident, incident or a fire is an essential part of the duties of every security officer. Every report must be accurate and clearly reported for ease of understanding, yet must be as concise as the circumstances being reported will allow.

Reports made by security officers will be used by management to correct discrepancies observed by the officer which pertain to security, safety and fire hazards, and suspicious activities and specific incidents, such as those referred to in previous chapters. In many cases the information which is gathered and reported by the security officer will become the basis for determining the cause of the incident. This knowledge will enable management to implement procedures which are designed to prevent future occurrences or which will provide evidence for later action or prosecution.

The security officer need not be a Hemingway or a Michener to be able to write a good report. It is not necessary that he be a walking dictionary or adept at working crossword puzzles. If he has a command of the spoken language, he can be taught to write a good report.

It does not matter if the report is for the recording of action taken or findings of a criminal investigation, provides the details of an employee's accident, pertains to the discovery of a fire or its later investigation, or is an account of an attempted pilferage; the principles and elements of the report are always the same. For the security officer the report is his method of communicating what he has observed and accomplished, during his shift, to his superiors in the security department and to management personnel of the firm that employs him. A report which merely states "all secure" following each patrol or for an entire duty tour is an indication that the officer is not alert and is not adequately performing his duties.

The investigator conducting a homicide investigation, the newspaper or television reporter covering a breaking story, and the author researching a book on an historical event all search for answers to certain questions. As they determine the answers, they record them for use in reporting the facts as they have learned them. The questions which must be answered by these people are the same that must be answered by the security officer when confronted by a security hazard that must be reported or when gathering information on a theft, fire or accident. Once the answers to these basic questions are determined, they form the elements essential to the writing of a good report.

The security supervisor is responsible for insuring that the members of the security force are properly trained as well as seeing that they are properly equipped. At each pre-shift inspection, it should be verified that each officer has a notebook and pencil in his possession. A security officer should no more be permitted to begin a shift without these items than he would if he reported in blue jeans and sneakers. Inside the front cover of each notebook, the essential elements of a thorough investigation and a complete report should be recorded. These elements are simple and easy to remember but should be recorded in the notebook since some of the answers needed might be overlooked during the hectic activity of an accident, fire or other emergency.

These elements are the answers to the following questions: **Who? What? Where? When? How?** and, whenever possible, **Why?**

Report writing is basic to the successful accomplishment of the security function. For this reason it is essential that the members of the security force be thoroughly trained in this subject. It should be a major part of their initial training and should be repeated as often as is required to insure that accurate and complete reports are being received concerning everything that takes place within the protected property. One of the major problems facing security operations is inadequate or incomplete reports being rendered by the security force.

Human nature being what it is, security officers tend to put into their reports only as much information as is insisted upon by their immediate supervisor. If that supervisor fails to review reports for content, fails to point out the deficiencies he finds, or fails to require that incomplete reports be corrected, the security department is going to end up with reports that consist of "all secure" as their entire content.

During the course of training sessions and the review of submitted reports, the supervisor must continue to stress that the six elements of a complete report be determined, written down in the officer's notebook and then included in his report of the incident. To assist in the training of the security force, a detailed listing is provided of what information should be obtained and included in every report submitted pertaining to discrepancies, accidents, fires or incidents.

WHO

Every person involved in an incident must be fully identified. This must include his full name, address or department, and social security or clock number.

This information must be obtained, recorded and reported on every person who has any connection to the incident being reported. This includes complainants, victims, suspects, witnesses and responding personnel from police, fire or medical departments, other security personnel who are involved, and management personnel to whom the incident is reported or who have responded to the scene of the incident.

WHAT

What happened? Was it a theft, accident, fire, safety violation or security hazard? You must determine, record and then report everything you can find out as to what actually took place. It should be stressed that the information which is desired is only that information which is essential to the accurate reporting of the incident.

If there are statements from witnesses, each witness must be identified in the report and a brief summary given of what each had to contribute. Written statements, if obtained, should be appended to the report. For instance, a witness has stated that he was eating his lunch shortly after noon, he remembered the time because his wife had forgotten to fill his thermos and he had to buy some liquid refreshment from the catering truck, to go with his corned beef on rye, when he heard and saw an explosion in the maintenance department. The report should reflect that witness Smith reported seeing and hearing an explosion in the maintenance department shortly after noon on the date in question.

WHEN

What time did the incident happen? When were you notified? What time did you arrive at the scene of the incident? What time did other persons who have been identified as responding arrive at the scene? What was the time when witnesses first became aware of the incident?

In recording and reporting the answers to the element "when", make sure that the hour, minute, day, month and year are included.

WHERE

In recording and reporting the location of an incident or accident, it is very important that the exact location be identified. If no landmarks exist at the location of the incident, with which it can be identified, then measurements should be made, from several known objects, to the location of the incident in order that it may be pin-pointed. If a theft, assault or incident took place in one location but was discovered in another, then both locations should be fully identified.

HOW

Determine, record and report how the incident, accident or violation happened. For example, "The employee slipped on an oil spot and was injured in the fall." "The radio was stolen from a broken carton which had been left unattended on the shipping dock." "The fire started in the paint shed *as a result of a spark from a grinding wheel.*"

WHY, IF POSSIBLE

Why did the incident happen? It is always important to determine the motivating factors and causes of an incident. In situations such as thefts, accidents, assaults or arsons, where the person responsible has not been identified, being able to establish a motive for the incident will often lead to the identification of the person(s) responsible.

In incidents where the security officer has been able to identify what caused the incident, the information will often be of great assistance to management in making policy changes designed to prevent similar incidents or accidents from occurring. The answer to the element "Why" may often be learned as it is determined how an incident happened, as stressed in the example above.

As the security supervisor continuously stresses to the security force the necessity of obtaining the essential information, recording it in their notebooks and including it in their reports, he will find that he is receiving all the information available, it has been accurately reported and the reports are easily understood.

There are various methods of writing reports so that they are concise yet provide the reader with an easily understood account of everything that took place. The report can start at the arrival of the reporting officer at the scene and merely report the incident as it unfolds for the officer. Many security departments and police agencies use this method of reporting. Another method, and the one preferred here, is reporting the incident in chronological order, as it happened as was determined by the reporting officer's information obtained at the scene and from witnesses. Either method will provide the reader of the report with accurate and complete information as long as all the elements of a good report are present.

The security supervisor should not stop with providing the members of the security force with notebooks, pencils and the knowledge necessary to write and submit a good report. There is one other aid to obtaining a good report which if implemented will greatly assist security officers in providing accurate reports of incidents which they have encountered. That is to provide the security force with preprinted forms designed to facilitate their reporting of different incidents which they might routinely encounter during the course of their daily duties. The type and content of these forms will vary depending upon the responsibilities given to the security officer, the nature of the business being protected and the frequency of certain categories of incidents.

On the pages that follow, some examples of these forms will be provided. They may be used as outlined or may be changed to suit the needs of a particular business. An example of a completed report submitted on an accident which has occurred on the protected property will also be provided.

Theft, accident and fire have been selected and an example of a security Shift Blotter is provided (Figure 23-1 to 23-6). In Chapter 17, a sample safety patrol report was provided which could be adapted to a routine patrol report, if desired. These reports have been selected because they should be adaptable to most security operations. Keep in mind that the simpler the form adopted

FIRM NAME
THEFT REPORT

Reporting Officer	Badge No.	T.R. No.	Date

Description of Property

Reported By	Victim (Person - Department)

Location (Fully Identify)	Estimate of Value

Details *(Continue on separate page, if required)*

Printed Name and Title of Reviewing Officer	Printed Name and Badge No. of Shift Supervisor

Reference Blotter Entry No.	Dated

Signature of Reporting Officer	Date Signed

Figure 23-1.   Sample form for reporting theft.

FIRM NAME			
FIRE REPORT			

Reporting Officer	Badge No.	F.R. No.	Date

Location of Fire (Fully Identify)

Reported By	Time of Report
	Time F.D. Notified
	Time F.D. Arrived

Estimate of Damage	Responding F.D. Supervisor

Details *(Continue on separate page if required)*

Printed Name and Title of Reviewing Officer	Printed Name and Badge No. of Shift Supervisor

Reference Blotter Entry No.	Dated

Signature of Reporting Officer	Date Signed

Figure 23-2.   Sample form for reporting fire.

FIRM NAME
SHIFT BLOTTER

Inclusive Hours		Date	
From	To		

Entry No.	Time	Reported By	Activity	Entered By

Figure 23-3.   Sample shift blotter.

for reports the easier it will be for the security officers to prepare. They will be able to provide a complete and accurate report of an incident while keeping the report as concise as possible. This combination should make for easier reading and understanding of the reports being submitted, and the time away from the job writing reports will be kept to a minimum.

The security supervisor or risk manager should always keep in mind that he may be supervising the most effective and efficient security operation in the nation, but in the absence of a good reporting system, no one is going to know of the job that is being done. It is also true that, in time, without a good reporting system, the entire security effort will begin to collapse. Losses cannot be reduced and risks cannot be managed without knowledge of what is taking place. That knowledge must come, in great measure, from the security reporting system.

FIRM NAME
ACCIDENT REPORT

Reporting Officer	Badge No.	A.R. No.	Date

Type of Accident (Personal - Vehicular - Property)

Reported By	Victim (Person - Department)

Location (Fully Identify)

Details (Continue on separate page if required)

Printed Name and Title of Reviewing Officer	Printed Name and Badge No. of Shift Supervisor
Reference Blotter Entry No.	Dated
Signature of Reporting Officer	Date signed

Figure 23-4.   Sample form for reporting an accident.

FIRM NAME
SHIFT BLOTTER

Inclusive Hours		Date
From	To	
4:00 P.M.	12:00 A.M.	SEPTEMBER 30, 1980

Entry No.	Time	Reported By	Activity	Entered By
25	6:05 P.M.	HARRIS	S.O. HARRIS REPORTED A MOTOR VEHICLE ACCIDENT IN THE NORTH PARKING LOT REF. A.R. #6	N.M.B.
26	6:30 P.M.	SMITH	SAFETY PATROL COMPLETED, S.R. #3 FILED	N.M.B.

Figure 23-5. Completed shift blotter report.

PATROL REPORTS

Reports from patrolling officers may be rendered as individual reports or may be incorporated into a shift blotter in which all activity taking place during the shift is recorded. Blotter entries are recorded by a dispatcher or a desk sergeant from information received from other security officers, employees and the general public by radio, telephone or in person. The blotter is the most thorough and efficient means of recording the activity of the security guard force. Cross referencing between blotter entries and report numbers is recommended. A blotter will usually take the form shown in Figure 23-5.

FIRM NAME
ACCIDENT REPORT

Reporting Officer	Badge No.	A.R. No.	Date
GEORGE P. HARRIS	807	6	SEPTEMBER 30, 1980

Type of Accident (Personal - Vehicular - Property)

SINGLE VEHICLE ACCIDENT

Reported By	Victim (Person - Department)
EDGAR S. JONES CLOCK NUMBER 1685 MAINTENANCE DEPT.	ALAN R. HAYS CLOCK NUMBER 385 PRODUCTION DEPT.

Location (Fully Identify)

200 FEET EAST OF GATE #1
NORTH PARKING LOT.

Details (Continue on separate page if required)

AT 6:00 P.M ON SEPT. 30, 1980 MR. ALAN R. HAYS ENTERED GATE #1 OF THE NORTH PARKING LOT, IN HIS PRIVATELY OWNED A 1980 FORD FAIRLANE LIC. # 367 LBN, NEVADA 1980, COMPANY DECAL NO. 385. MR. HAYS STATED THE BRAKES ON HIS AUTO FAILED CAUSING HIM TO COLLIDE WITH A FENCE POST LOCATED 200 FEET EAST OF GATE #1.

THERE WERE NO SKID MARKS AT THE SCENE. A CHECK OF BRAKES OF THE VEHICLE INVOLVED INDICATED THEY WERE INOPERATIVE. THERE WERE NO INJURIES AND NO ALCOHOL INVOLVED. PROPERTY DAMAGE ESTIMATED AT $350.00.

Printed Name and Title of Reviewing Officer	Printed Name and Badge No. of Shift Supervisor
ALZON A. REED SUPERVISOR	NILA MARIE BIEKER NO. T2277

Reference Blotter Entry No.	Dated
25	SEPTEMBER 30, 1980

Signature of Reporting Officer	Date signed
Alzon A. Reed	*October 1, 1980*

Figure 23-6.   Completed accident report.

QUESTIONS FOR DISCUSSION

1. What are some problems in relying solely on the spoken form of communication?
2. Does expressing observations made in writing require a college degree?
3. Why are written reports important to the effectiveness of the security program?
4. Do the elements of a thorough report vary depending on the type of report being written?
5. Will reporting "all secure" provide the necessary information to enable management to evaluate the security officer or the security program?
6. What information is required to be given about persons involved in an incident being reported?
7. How important are "where and when" to personnel who are responsible for evaluating an incident or determining if prosecution is called for?
8. In reporting what happened, should statements from witnesses be included in a report? In detail? In summary?
9. Is it possible to determine motive or cause from finding out "how" an incident happened?
10. Why is it important to determine motive or cause of the incident being reported?

MAJOR CONSIDERATIONS

1. A clear, concise, complete report of security force activity is necessary if a valid security program is to be maintained.
2. Oral communication tends to be changed when being passed from one person to another due to poor memory, exaggeration or personal perceptions.
3. Standard report forms make report writing easier, less time consuming and more uniform.

Chapter 24

Public Relations

When most of us think of public relations or P.R. men, we tend to picture the Madison Avenue type pitchman or ad person working in the glamorous world of business executives, models, movie stars and high society. Actually, anyone who must relate to the general public during the daily performance of his work is required to practice public relations.

For the security officer, public relations is an important and full-time part of his day-to-day duties. The involvement of the security officer in public relations begins as he is dressing to go on duty and ends upon his return home following his duty tour.

APPEARANCE OF THE GUARD

Since the security officer is one of the first representatives of his employer's business that a visitor to the business will encounter, the appearance of the officer will go a long way in determining the visitor's ultimate opinion of that business.

If the officer's appearance is sloppy, disheveled or even unkempt, the opinion of the visitor is likely to be that this probably typifies the attitude and management of the business. On the other hand, if the security officer is clean-cut and neat, the visitor's first impression of the company will most likely be that the company is efficient and well managed.

To insure that a guard force presents a favorable first impression to visitors to the facility, standards must be set for dress and appearance. These standards should be reduced to written form and given to each member of the security force, preferably on the first day of their employment. During their initial briefing they should be advised that the standards will be strictly and impartialy enforced, must be adhered to and that failure to maintain the standards will result in appropriate disciplinary action.

A standard for disciplinary action should also be reduced to writing, and both standards, along with other appropriate rules, should be made part of the security force's work rules. An example of appropriate disciplinary action for violation of dress or appearance standards would be as follows:

1. Minor infraction—Written reprimand placed in employee's personnel file.
2. Substantial infraction—Employee not accepted for duty and sent home without pay.
3. Flagrant or repeated infractions—Termination of employment.

Having established standards and provided for disciplinary action, the standards must be enforced. Inspections should be conducted at briefing sessions, prior to the assumption of duty by each shift, each day. Shift supervisors should be charged with responsibility for insuring that their personnel meet the standards prior to going on duty and continue to maintain the standards throughout the shift.

It should be further required that the standards be maintained at all times that guard force personnel are in the public view, in uniform. This would include traveling to and from their workplace. Particular emphasis should be placed on a prohition of partaking of alcoholic beverages in any public place while wearing any recognizable part of the uniform. At the end of this chapter, a sample standard for dress and appearance is provided, which can be used as a guide in the preparation of standards.

APPEARANCE OF THE GATE OR GUARD HOUSE

Following the security officer, the next opinion-forming impression that a visitor to the business will receive comes from the area under the immediate control of the security officer—the gate house, guard house, guard office or other area occupied by a member of the security force. Once again, standards should be prepared outlining the responsibilities of the guard force in maintaining a high standard of good housekeeping within these areas. This standard should include all aspects of good housekeeping including littering, dusting, trash removal, sweeping and any other factors affecting the appearance of the area. It should also list those departments, accompanied by the appropriate telephone number, which are responsible for necessary repairs and major housekeeping duties such as window washing, floor waxing and painting.

DEMEANOR

The next area of public relations with which a guard force must be concerned is that of the individual officer's demeanor while in contact with the public. How the security officer greets and conducts himself while dealing with visitors or other employees of the company has a direct bearing on the public image and opinion of the company.

Surliness, abruptness, heavy-handedness, or obnoxious treatment are not signs of macho manlihood or of liberated womanhood. They are signs of ill-mannered, unprofessional conduct. The officer who handles authority coolly, being courteous, even when firmness is required, is the true professional.

Courtesy while signing in or directing visitors, checking identification cards of employees, answering and conversing on the telephone, dealing with a pilferer or trespasser, or while involved in an emergency situation marks the security officer as a capable, confident individual and does wonders for the image of the company which employs him. Telephone demeanor is of particular importance since many contacts with a business are made solely by telephone. To make sure that a good image is being projected, a standard should be established for the proper method of answering and conversing on the telephone.

The knowledge or lack of knowledge which the security officer displays in his daily performance of duty also has a bearing on the formation of public opinion. The security supervisor should make sure that all of his guard force personnel are well trained in all aspects of their duties. Their ability to answer questions and direct visitors should be tested and monitored. They should be furnished with diagrams, with routes to various areas plainly marked, to give to visitors, thus enabling the officer to provide accurate information and assistance.

TRAINING AND MOTIVATION

Public relations for the security officer boils down to pride in his personal appearance, in the job he performs and in the area in which he works. It includes the knowledge he has gained and which he displays in his dealings with company employees, visitors to the area and management personnel. It is further reflected by the courteous manner in which he performs his daily routine. In short, it is professionalism.

The success or failure of the public relations effort put forth by the security force will ultimately be decided by how well their supervisor has motivated, trained and supervised the security forces who work for him.

There are some particular assignments where the need for public relations is even more pronounced. These include those assignments which bring the security officers into direct contact with the general public throughout the course of their daily duties. Such assignments as hotels/motels, casinos, airports, retail stores and shopping centers, banks, sporting events, conventions and other special events require a higher degree of tact and courtesy on the part of the security officer.

In the typical industrial setting, the guard force is dealing primarily with company employees. Visitors to the protected area basically fall into the category of salesman and service personnel who are vying for the business of the guard force's employer. Although a negative image or opinion of the guard force is never good, it is not as damaging under these circumstances as a negative opinion would be of a guard force assigned to a business or facility engaged in serving or selling to the general public.

It is doubly difficult, but absolutely necessary, that while in constant contact with the paying guests or customers, the rules be enforced and the property of the employer be protected in a courteous and tactful manner. In supervising a security force under these circumstances, it should be stressed to all personnel that although their primary function may be to protect the assets of their employer, they have a secondary duty to provide protection and assistance to the customers and guests of their employer.

It will never be easy to be courteous to the obnoxiously intoxicated person, the chronic complainer or the know-it-all, but the job will be more pleasant and will be made much easier when each task is performed in a highly professional manner.

Nowhere is the admonition "Do unto others as you would have them do unto you" more appropriate than when your livelihood and that of your employer is dependent upon the respect and good will of the customer. This is not to imply that the thief is to be ignored or that the intoxicated or trouble maker given carte blanche. However, even these situations can be handled with tact and courtesy. To do so will reflect favorably on the guard force, its supervisor and their employer in the eyes of the visitor, guest or customer.

You can tell a professional by looking and listening. Look around you and decide what type guard force are you supervising? If it is not what you want, you can improve it through training, guidance and supervision.

UNIFORM AND APPEARANCE STANDARDS

Each member of the security force is expected to maintain the highest standards of personal hygiene and appearance. Security officers are required to wear a neat, clean and well-pressed uniform. Shoes will be in good repair and shined to a luster. It is the responsibility of the shift supervisor to provide the example and to inspect each officer assigned to his shift to insure that all officers of this force comply with these standards.

Inspections will be conducted daily prior to the assumption of duty by each shift. Failure to be in the prescribed uniform when reporting for duty or while on duty will be grounds for the imposition of appropriate disciplinary action as outlined elsewhere in the standard operating procedures for uniformed personnel.

The personal grooming of security force personnel must not detract from the proper wearing of the uniform and equipment. Security officers are required to present a neat appearance and be consistent with the following standards.

Male Personnel

1. The face will be kept clean shaven except that a mustache may be worn if desired by the individual officer. If a mustache is worn, it will be neatly trimmed.
2. Hair will be neatly trimmed at all times and will be clipped on the sides and back to present an evenly graduated appearance. Hair will not be worn below the level of the collar.
3. Hair on top of the head will be neatly combed, brushed or styled in such a manner that it does not extend past the band of the cap.
4. Sideburns will be neatly trimmed and will not extend past the lower edge of the ear nor be wider than one and one-half inches at the bottom.

Female Personnel

1. Hair will be worn so as to be close to the head and not extend below the top of the uniform collar.
2. Hair may be kept longer than prescribed above only if it can be swept up on top of the head and does not interfere with the proper wearing of the issued head apparel.

Uniform Wear

Standards for uniform wear must include all items of the uniform that are required to be worn and a prescribed manner in which each item is to be worn must be designated. Normal items of uniform apparel will be as follows:

Summer.   Short-sleeve shirt, trousers or slacks, belt, shoes, socks, badge, hat or cap, hat or cap insignia, tie (optional).

Winter.   Jacket, long-sleeve shirt, tie, trousers or slacks, belt, shoes, socks, hat or cap, hat or cap insignia.

In the event there are major differences in male and female uniforms, they should be listed separately. Chevrons or other types of rank insignia should be issued and the manner in which they are to be worn designated.

In the event all items of the uniform are issued with the exception of shoes and socks, it should be clearly designated what style and color of shoes and socks are acceptable.

If uniforms are to be purchased by members of the guard force, the color, style and type of material which will be acceptable must be designated.

QUESTIONS FOR DISCUSSION

1.   What is public relations?
2.   Why should a security officer be concerned with public relations?
3.   If a security officer's appearance is unkempt, what impression is given of the company that employs him?
4.   Why would the appearance of the guard house be important to good public relations?
5.   Is it possible for a security officer to enforce company rules in a courteous manner?
6.   What is the mark of the professional security officer in his dealing with people he encounters?
7.   As a professional security officer, is it possible that what you do not know can hurt you?
8.   What are some types of security assignments in which public relations takes on even more importance? Why?
9.   What duty does the security officer owe to the customers of his employer?
10. How can the image of the security guard force be improved?

MAJOR CONSIDERATIONS

1.   The appearance of the security officer and the area to which he is assigned are major concerns in establishing a good public relations program.
2.   In the professional world of security, courtesy and knowledge are two of our most important products.
3.   Public relations is always an important part of a viable security system but especially where the security officer is constantly dealing with the general public.

Chapter 25

Firearms

Whether security officers should be armed is probably the most controversial of all the questions raised in planning or implementing a security guard force. The answer is, of course, very simple: yes and no. Each and every position filled by a security officer must be examined separately to determine if the officer should be armed and, if so, with what type of weapon. There are many types of weapons from which to choose, but the final choice is usually an eye irritant, night stick, hand gun or, in special circumstances, shotgun.

Many factors come into play in determining if a guard force or even selected members of a guard force should be armed. Some of the factors to be considered are as follows:

1. Does the property, product or person being protected justify the taking of a life to maintain their security?
2. Is the performance of the officer's duty so inherently dangerous that he needs to be armed for his own self-protection?
3. Is the protected property located in a high crime area?
4. Does the liability insurance carrier believe the position should be armed?
5. Is the police protection so thinly spread that their response cannot be depended upon in the event of an emergency?
6. Is there a steady work force or does the industry have heavy turnover and rely primarily on transients?
7. Are background checks conducted before hiring?
8. What has been the experience of like industries in similar areas as pertains to arming security officers?
9. Would the use of firearms be more dangerous and costly in potential loss of life and property than not having them available for use?

10. Have state and local laws pertaining to the arming of security
     officers been researched? Will the guard force be permitted to be
     armed and can existing members of the guard force qualify to carry
     weapons under those laws?

The decision should be individually made for each position on
the guard force. The person or group which would be responsible
for any accident, misuse or illegal act on the part of the potentially
armed guard force should be the decision maker on this question.

Once the decision has been reached and if select members of
the guard force, or the entire guard force, is to be armed with the
deadly force of firearms, some reorganization of the guard force
may be necessary. Keep in mind that the position should determine
the qualifications and the capabilities of the person filling the
position. The horse should never be put before the cart under these
conditions

Remember good old Joe? He has been a member of the guard
force for five years now. He is clean, neat and you can depend
upon his being there when he is scheduled. Joe isn't too bright and
tends to be a mite jumpy when he is alone at night on the back door
leading to the parking area. He probably doesn't need to be armed
but we didn't want to hurt his feelings by letting everyone else have
a gun and not give him one. It sure is too bad about that nice Vice
President of Marketing who worked late the other night and
startled good old Joe, who promptly shot him.

Far fetched? Unfortunatley, it is not. There are security
officers who are deaf, cannot see over five feet away and are totally
illiterate who carry a hog leg strapped to them which would have
made Wyatt Earp envious. The logic seems to go as follows: We
need guards but we don't want to pay any more for them than we
have to, so we will hire whatever we can get and make up for what
they lack by giving them a gun.

The statistics on security officers who are shot in the line of
duty indicate that a good many of them are shot with their own
weapon which had been taken from them. In some of these cases
the intruder(s) did not have a gun when they entered, but they had
one when they left.

SELECTION

Security officers who are to be armed should have more stringent hiring and
retention standards than anyone else employed. After all, other employees can

be honest, great at their job and still be a little flakey. This cannot be so with an armed security officer. Before placing a gun and the inherent authority to use it in the hands of any individual, his moral, physical and psychological make up must be unquestionably sound.

Some employers have applied a different solution to the problem. They issued hand guns without ammunition and forbade the guard force to load the weapons. They believed that the mere presence of the gun would be a sufficient deterrent. Of course, they may not have realized that they were setting up their guard force as clay pigeons for any psycho or hop head that believed they had to fight force with more force or at least equal force.

Ideally, hand guns carried by a guard force should be provided by the employer, maintained by the employer, serviced by the employer and when not in use they should be secured on the property of the employer. The same is true of ammunition for the weapons.

Doing this may be a bit more costly initially, but the cost will more than be justified and offset with the passage of time. The problem of possible defective weapons or ammunition, the possibility of a weapon being stolen from the home or auto of a guard and used in criminal acts, or the possibility of accidental discharge or use to commit suicide or homicide during family or neighborhood disputes will have been eliminated.

There was a gentleman who had been a tennis professional and had been unable to contiue in that profession due to physical condition and increasing age. He turned to the security field to earn his living and had been assigned to a particularly tough assignment in a high crime rate area. It was later determined that he had neither the experience nor the ability to handle the problems he would encounter on this particular assignment and he was relieved to be reassigned the following day. His lady friend picked that time to decide that they were no longer compatible. That night, while laying in his bed in his rented room, he took his issued revolver, placed it to his head and ended his life.

Whether weapons are issued or security officers are permitted to use their own, it should be required that the weapons be standardized and inspected regularly. The inspection should include checking the ammunition for loose projectiles and cleanliness. Ammunition should be clean but should not be shined or oiled.

Saturday night specials, the cheap inferiorly constructed hand guns, should be ruled out immediately. They might cost less initially, but the liability assumed if they blow up in the guard's hand or are so inaccurate that they hit everywhere but where they are aimed, not to mention more costly repair bills and the shorter period of usefullness will undoubtedly cost much more in the long run.

The heavy magnum weapons, whether .357, .44 or .45, should not be considered for use as hand guns. They are much too powerful using magnum ammunition. Although the .38 cartridge may be used in the .357, it is not recommended that this be done unless ammunition is to be inspected daily and perhaps even several times a day to insure that personnel are not using their own loads.

The most common hand guns used by police departments or security guard forces are .38 caliber military and police special revolvers with a standard 4″ barrel. The best manufacturers of these guns are Smith and Wesson and Colt.

If semi-automatic pistols are preferred, both Smith and Wesson and Colt manufacture good .38 and .45 caliber models. The most popular over the years has been the Government model .45 caliber, M1911A1 pistol manufactured by Colt. This weapon was designed in 1905, improved and adopted in 1911 and, except for a few minor improvements, it is the same reliable weapon used today by army military police, tankers and officers.

TRAINING

Weapons training can be broken down into four major categories: safety, care and handling, firing of the weapon, and restriction on use of the weapon.

Safety

Accidental discharge of firearms results in countless injuries and deaths each year. These accidents are caused due to lack of knowledge, failure to properly secure, careless handling, or using the deadly weapon as if it were a toy. It should go without saying that safety in the handling and use of a weapon must play a major role in the training of anyone who is to be issued or carrying a firearm. Safety instruction must be carried a step further and be provided for husbands, wives and children of security officers when the weapon is to be carried in the family car or stored in the home when not in use. To insure safe handling of a weapon, the following safety precautions should be followed.

1. Always consider a weapon to be loaded whenever it is handed to you or you pick it up. Be sure that you check it *each time* to make sure that it is not loaded. Never allow anyone in your presence to handle a weapon without knowing how to check or without checking the weapon itself to see that it is not loaded. This should pertain to family members, co-workers or subordinates.

2. Even though you know the weapon to be empty, never point it at anyone, whether in jest or by thoughtlessly waving it around during conversation. Remember the cardinal rule of handling firearms:

Never point a weapon at anyone unless you are prepared to take his life. Following this rule explicitly and insuring that anyone around you handling a weapon also adheres to this rule will result in the prevention of accidental shootings.

3. When storing a weapon, unless it is secured in a location where it may have to be ready for immediate use, always unload the weapon and store the ammunition in a separate, equally secure location.

4. Whenever handling a weapon on a dry or live firing range, check it each time you pick it up. While you are handling it, loaded or not, always keep the muzzle pointed down range in the direction of the targets. After a firing exercise *always* unload the weapon before turning away from the firing point or putting the weapon down.

Care and Handling

Major disassembly or repairs of a weapon should be done by a gunsmith or a qualified small arms repairman. A weapon should be thoroughly cleaned after firing, using a good solvent. After cleaning it should be lightly oiled. The weapon should then be cleaned daily until the cleaning patch used on the bore no longer picks up gun powder residue. Hoppes No. 9 has a good reputation with gun users for providing quality cleaning solutions, rods, patches, oils and so forth.

Weapons which are not being fired should be cleaned several times a week. In hot humid areas the weapons should be wiped clean after each use and lightly oiled to prevent rusting.

Every person who is required to carry a deadly weapon should strive to become an expert marksman with the weapon. If the situation arises where it is necessary to draw and fire a weapon, it is obviously important that no innocent bystander be injured because the accuracy of the person firing the weapon was not what it should be.

Expert marksmen are born in the classroom, developed on the dry firing or practice range and only prove their expertise on the live firing range. The basics of accurate fire such as position, grip, sight, alignment, area of aim, breath control and trigger squeeze must be learned and then practiced until the shooter performs them correctly without being consciously aware of them. It is recommended that security officers who have not used hand guns before or who have had problems in their use be taught a basic marksmanship course consisting of firing on bullseye targets. Once this has been mastered, they will be ready to advance to the more difficult practical police course or police combat course.

Position. To obtain a good position for a basic marksmanship course, a right-handed shooter should squarely face the target, turn 90° to his left and raise his right arm in the direction of the target. With the arm extended, the hand should be loosely closed as if it held a weapon. The shooter should then sight down the arm over the hand. The hand should be positioned squarely in the center of the target. If it is not centered, the shooter must not try to bring it to bear on the center of the target, but should drop the arm, look away and adjust his position by moving the left foot slightly in the opposite direction of that needed to bring his arm into the center of the target. This exercise should be repeated until the arm when raised is exactly centered on the target with no adjustments. For a left-handed shooter the procedure would be just the opposite as described above. This whole exercise, after practice, should only take a few seconds. However, keep in mind that in firing a markmanship course the presumption is that the target will not be firing back.

The purpose of attaining a firing position in this manner is that once accomplished the arm is in the most natural and relaxed position possible. It will return to that position after each round has been fired and the shooter will not have to be continually moving the weapon back to the center of the target. This also results in reducing the strain on the shooter's arm muscles and allows him to hold the weapon steady.

A good position for firing the police combat or practical police course depends upon the course being fired and the positions selected for use. Possible positions are from the crouch, standing, kneeling, sitting, prone and shooting from behind a barricade.

To obtain a good crouch position for snap shooting from the hip, the shooter should face the target, extend the left foot (right-handed shooter) slightly forward, and bend at the knees, keeping the upper part of the body perfectly straight. While assuming the crouch position the weapon is drawn. The upper portion of the arm is kept tightly against the side. The weapon is brought up to bear on the target by bending the arm at the elbow. This position is used for firing on silhouette targets from about seven yards distance. This distance was selected since statistics show that most gun fights occur when the antagonists are from five to ten feet apart. When shooting from this position there is a tendency to shoot low. Care should be taken to avoid this problem by insuring that the arm is raised slightly above the perpendicular.

To obtain a good crouch position for snap shooting using the two-handed grip, the shooter should have the legs apart separated by about the width of his shoulders. He should bend at the knees, lowering the body. The upper portion of the body must remain straight. The weapon is drawn and extended to arms length toward the target. The other hand is brought forward and can be placed beneath the shooting hand for steadying the weapon or wrapped around the hand holding the weapon. The two-handed grip is preferred to the hip shooting method, time permitting. The difference is only a few seconds, but since in some situations that length of time may not be available, both methods should be learned and mastered. In using this stance the weapon should be held at eye level and directed toward the center of the target.

Other positions to be used in firing the police courses, such as kneeling, sitting and prone, are more flexible and can be adapted to the individual shooter. The basic requirements are that the position used minimizes the exposed area of the shooter and that it provides a steady platform for the weapon. The shooter should avoid bone-to-bone contact such as resting the elbow on the knee. In this situation there is a tendency for the arm to rock back and forth, depriving the shooter of the steadiness desired. When kneeling, the shooter should try to have the left upper arm as far over the left knee as possible to provide support for the firing hand. In using the sitting position the arms can be tightly enclosed between the inside of the legs in such a manner as to provide good support.

When shooting from behind a barricade, the shooter must master shooting with either the right or left hand to insure that he is able to keep as much of the body as possible protected by the barricade rather than exposed to hostile fire. Barricade shooting may be done from the standing, kneeling or prone position.

Grip. A good grip prevents the weapon from moving within the hand and permits the shooter, once on target, to stay on target with a minimum of effort. When drawing the weapon the web between the thumb and forefinger should be centered on the stock. The grip should be obtained by the remaining three fingers which are wrapped around the stock tightly but without squeezing. The position of the thumb varies depending mostly on individual preference. The consensus is that when using a pistol the thumb should be up along the slide, while with a revolver it should be curled down toward the trigger guard.

Sight Alignment. Correct sight alignment is obtained by centering the front sight blade between the two upright portions of the open rear sight and keeping the front sight blade level with the upright portions of the rear sight.

Area of Aim. Traditionally, when firing a markmanship course the sight picture was obtained by aligning the sights and then sitting the base of the bullseye centered atop the front sight blade. With a correct sight picture the rear sight and bullseye should appear slightly hazy to the eye, while the front sight blade would be sharply in focus. Using this sight picture, trigger squeeze was stopped anytime the picture changed and pressure was held on the trigger but not increased until the correct sight picture was again obtained. Experience has shown that for many shooters, holding at the base of the bullseye causes a strain and a resultant tendency to jerk the trigger before the weapon moves.

This problem can be overcome by giving the shooter an area at which to aim and allowing him to continue to squeeze the trigger as long as the weapon remains anywhere within the area designated. Proper sight alignment and trigger squeeze and aiming at an area covering the lower third of the bullseye and extending out to the eight ring on either side results in the shot impacting within the area of aim. In firing the police practical or combat courses using the silhouette target, the area of aim is taken in the center of mass of the target. This would be just above and in line with the belt buckle on the average man.

Breath control.   During firing, the control of breathing must be practiced to prevent the weapon from rising and falling with inhaling and exhaling. This is accomplished by obtaining good sight alignment and taking a deep breath once in the area of aim, expelling half of it and holding the rest until after the weapon is fired.

Trigger squeeze.   Accurate fire can be brought to bear on the target only by incorporating all the principles of good marksmanship. Of the stated principles, the most important for building and developing skills are sight alignment and trigger squeeze.

Trigger squeeze can best be obtained if the shooter can use the pad of the first digit of the index finger on the center of the trigger. This is not always possible due to the variances of hand versus gun size. To obtain a good trigger squeeze, pressure should be applied to the trigger, directly to the rear with a constant increase in pressure until the weapon fires. With correct trigger squeeze, the shot will come as a surprise to the shooter.

Range Firing

If the opportunity arises to construct a range, plans for constructing indoor or outdoor ranges can be purchased, for a very nominal fee, from the National Rifle Association (NRA) in Washington, D.C. The NRA also has training aids to assist in conducting weapons training. It is recommended that the NRA be contacted for a list of available materials, films, range plans, etc.

Basic Markmanship Course.   A good basic marksmanship course is conducted from the 15, 25 and 50 yard lines with all firing conducted from the standing position. Firing is accomplished using bullseye targets and scored by adding the value of each hit. Individual qualification standards can be set unless there are state requirements. A good rule of thumb to use, however, would be that a shooter attaining 70% of the possible score (number of rounds fired times 10) can be classified as a marksman, 80% would be a sharpshooter and 85% or more would qualify as an expert.

This course has no value as a practical course in shooting for anyone engaged in law enforcement or security. However, it has excellent value in allowing the shooter to learn the basics of good markmanship, allowing him to become familiar with the weapon he is using and, most importantly, building his confidence in his ability to bring accurate fire to bear on a target.

Practical Police or Police Combat Courses. These courses are conducted from 7, 15, 30 and 50 yards. Each course of fire is started with the weapon holstered. Upon receiving the command to fire, or when the targets face forward, the position is assumed while the weapon is being drawn and firing commences immediately as each shooter has properly brought his weapon to bear upon the target.

From 7 yards, firing either from the hip or with the arms extended using the two-handed hold, the weapon is fired from the crouch position. Snap shooting does not imply that the trigger is being snapped or jerked, but merely that time is not being taken to align the sights of the weapon. The weapon is directed at the target in the same way as pointing your finger at an object. The weapon is merely an extension of the pointed hand. Using this method, trigger squeeze takes on added importance since the shot can be pushed or pulled off the target if an improper squeeze or jerk occurs.

Firing from 15 yards is conducted from the standing position with the arms extended fully using a two-handed grip. Again, the weapon is pointed at the target without taking the time to align the sights.

Firing from 30 yards is conducted from the standing and either the kneeling or sitting position. From this distance, sight alignment and area of aim must be used and combined with proper trigger squeeze to bring accurate fire on the target.

From the 30 yard line, firing is conducted shooting right handed from the right side of a barricade and left handed from the left side of the barricade. Positions can be designated or left to the option of each shooter.

Prone position fire is conducted from the 50 yard line. Use of a steady firing platform, good sight alignment, area of aim and proper trigger squeeze becomes more important as the distance from the shooter to the target increases. The positions are designed to minimize the shooter as a target while allowing him to bring accurate fire to bear. Any cover that an officer can find while engaged in a fire fight should be used to advantage. To accomplish this the shooter must be able to fire accurately from whatever cover is available. Shooting is a sport for some people, but for the security officer it is a deadly serious matter for which he must be totally prepared for the protection of his own life and the lives of those he has been hired to protect.

Restriction on the Use of Firearms

For most security officers there are only two sets of circumstances when they are permitted under law to use their firearms.

    1.   The firearm may be used by the officer to save his own life.
    2.   The firearm may be used by the officer to save the life of another.

It should be remembered that any use of the weapon will result in an investigation into the facts and circumstances surrounding the shooting. If the investigation indicates that the shooting was not justified, criminal charges can be brought against the security officer.

In a situation where the security officer must decide if he should use his weapon, he should keep in mind that the weapon is not to be used to bluff or intimidate. An officer who draws his weapon as a threat may find himself forced to use the weapon or loose it. Under these circumstances the investigation and subsequent court action might find that the guard was guilty of manslaughter or even murder. The burden of proof will be on the officer to show that he was justified in using his weapon—that his life or someone else's life was in danger.

The only exception to this restriction is if the officer has been given a police commission or has been deputized with the accompanying authority to act as a police officer. In these cases the officer is also authorized to use his weapon to prevent the commission of a felony or to prevent the escape of a felon. Many states have specific laws governing the carrying and use of firearms by security officers. These impose minimum training standards coupled with minimum requirements to be met by the officer in the proficient use of his weapon as demonstrated on the range.

The State of South Carolina has set such standards, but once they have been met, the security officer while on duty has the same authority as any police officer in the state. The weapon may be carried to and from the place of employment but the officer is not permitted to make any stops whatsoever while en route. This stipulation is never waived and does not permit the officer to stop for coffee, conversation or any other reason while he is armed. If he does stop, he is subject to arrest.

Other states, such as Texas, do not permit the carrying of a firearm into any place that serves alcoholic beverages. Texas, Georgia and Illinois, to name a few, are some of the other states requiring minimum standards for training and proficiency before certifying an officer to be armed. These standards are not limited to firearms training and include knowledge of other security subjects as well as local and state laws.

The security officer who carries his weapon from his workplace into public areas not only has a responsibility to his employer, but also must be responsible to the people of the state where he resides. The employer who certifies a guard to be armed must assume responsibilities and liabilities for the actions of that officer. The decision to arm a security officer is a serious one and should not be made lightly without full consideration of the ramifications and liabilities involved. The information provided here should allow that decision to be made based on sound principle.

QUESTIONS FOR DISCUSSION

1. Discuss the factors to be considered before determining the necessity of arming the security guard force.
2. Who should be responsible for making the decision to provide weapons to the guard force?
3. Is it necessary to arm every member of the guard force?
4. Should it be necessary to adopt a more stringent selection process for armed guards than those used in the hiring of other employees?
5. What is the best method of insuring that safe weapons are used by members of the guard force while eliminating the risk of theft or accidental discharge of the weapons in the possession of off-duty security personnel?
6. Discuss the elements necessary in establishing a good weapons training program.
7. Can any benefits be derived by providing a basic markmanship course for the guard force?
8. Can it be assumed that an expert marksman will safely handle his weapon at all times?
9. What are the major restrictions imposed on armed guards?
10. What are the two circumstances when a security officer will be considered justified in using deadly force?

MAJOR CONSIDERATIONS

1. Only those security officers whose positions could require the use of deadly force should be considered for being given the authority to bear arms.
2. Weapons used by members of the guard force should be provided by the employer or the employer should exercise strict control over the type of weapon authorized for use.
3. Security officers who are to be authorized to carry firearms in the course of their duties should be more carefully selected, more

thoroughly trained and more closely supervised than other employees.

V. LEGAL CONSIDERATIONS AND TRAINING REQUIREMENTS

Chapter 26

Legal Authority For Security Personnel

There is a great deal of controversy among security personnel concerning the separation of security from the public safety agencies. On one side it is believed that no comparisons should be made and that the term private police should be expunged from our vocabulary. It is the belief of this segment of the security profession that by denying the connection and minimizing police-like duties, security personnel are more apt to receive recognition as risk managers and professional preventers of losses.

The other side of the controversy believes, as does the author, that it is impossible to separate security from a degree of law enforcement or to separate loss prevention from crime prevention. Even if it were possible to eradicate the joint history of public and private safety and security operations, it would be a mistake to do so. All previous expertise in the protection of assets through crime prevention must be maintained and built upon using this experience as a solid base for developing all the skills necessary to become viable risk managers. All professional fields are constantly changing and searching for better methods and procedures for improving performance. Security is no exception

Those in the security industry who believe that security personnel must divorce themselves from the law enforcement image will disagree with the extent of coverage provided in this chapter to define and explain criminal laws. Failing to fully acquaint security officers with the law as well as their responsibilities under the law amounts to failure in assuming the responsibility of risk management.

Security officers who are not totally aware of the law can be equated with the electrician who has been trained in all aspects of electrical wiring and repair except when and how to turn off the power. They are a definite loss risk. Some of the risks which may be incurred due to a failure to fully understand the law are civil actions against the employer for false arrest, false imprisonment, unlawful search and seizure and wrongful death.

The material and information presented here are essential for a security officer, and this information is presented in such a manner so as to permit the security officer to know what lawful actions he may take and what actions are denied to him under the law.

The security officer is employed to protect the property of his employer and to insure the well being of personnel who work at or visit the property. The duties of the security officer while within the protected area are not at all unlike the duties ascribed to the public law enforcement officer. The security officer must enforce company rules and regulations; his duties require that he prevent the theft of company property; he must prevent, respond to and report accidents; he must conduct spot checks of empoyees' personal property and vehicles; he must conduct patrols of the protected area; he must respond to and handle emergency situations; he must safeguard life and property from injury or damage, and he is required to submit reports of his activities. However, except when special authority has been granted in the form of a commission or deputation, the security officer is not a legally appointed public law enforcement officer. He has no more authority to make an arrest or to conduct searches than has any other private citizen.

A private citizen, ergo a security officer, may make an arrest for a misdemeanor which has been committed in his presence. A citizen may make an arrest when he knows for a fact that a felony has been committed and has reasonable cause to believe that the person being arrested committed the felony. In both of the cited circumstances the arrest is being made without a warrant.

It should be pointed out that the law cited here is common law, and although common law has been used as the basis for most of our statutes in this country, the law differs from place to place. The security supervisor or risk manager should obtain copies of the applicable law in the jurisdiction where he is employed and make sure that his security force personnel are thoroughly familiar with the provisions of those laws.

A legally appointed law enforcement officer may make a felony arrest, without a warrant, where he has reasonable cause to believe that a felony has been committed and the person being arrested had committed the felony. For the protection of the officers and the employer, it is essential to point out the very great difference in the authority to make a felony arrest without a warrant. The security officer *must know* that a felony was in fact committed while the law enforcement officer *must only reasonably believe* that a felony was committed.

Problems have been encountered in the part-time employment of off-duty law enforcement officers or in the employment of former law enforcement officers. The employer believes that the training provided the public officer is superior, rather than different from that given to a security officer and does not see the need for additional training of such personnel. When this officer is confronted with an arrest situation, he will respond based on his knowledge of the authority vested in a public law enforcement officer, which may exceed the authority to act as a private citizen, giving rise to a false arrest suit. The penalty for error in such cases can be severe for both the officer and the employer, since both may be subjects of civil suits for false arrest or imprisonment. The officer may even be subject to criminal charges stemming from the incident.

Only a duly authorized, appointed public law enforcement officer may make an arrest pursuant to the issuance of an arrest warrant. If a security officer *knew* that a felony had been committed and if he knew that warrant had been issued for the arrest of a certain individual, he might be justified in arresting that person based on his factual knowledge of the felony and his reasonable belief that this individual committed the known felony. He could not, however, make an arrest based on the issuance of the warrant alone.

In most jurisdictions, neither the public officer nor the security officer is authorized to make an arrest for a misdemeanor which has not been committed in his presence. The public officer may make an arrest for a misdemeanor not committed in his presence, following the issuance of a complaint and an arrest warrant. The security officer *may never* make an arrest for a misdemeanor which has not been committed in his presence.

Some jurisdictions specify, by statute, which criminal acts constitute a misdemeanor and which constitute a felony. Most, however, only specify the maximum penalties which may be imposed for each type of offense and allow the maximum penalty to determine the classification of the crime.

In most jurisdictions, a misdemeanor is classified as any violation of the criminal code for which the maximum penalty that may be imposed is a year's confinement or less. A felony is classified as any violation of the criminal code for which the maximum penalty that may be imposed is more than a year's confinement. Some jurisdictions determine the classification of a crime by the place where the confinement has been ordered. In most cases the end result is the same. For instance, if the sentence is to be served in a county jail or work farm, the crime is classed as a misdemeanor. If, on the other hand, the confinement is ordered to be served in a state penal institution, then the crime is classed as a felony. Confinements are never made in state penal institution in cases where the sentence is for less than a year's confinement and rarely, if ever, are made in a county penal facility for more than a year. An exception could be the imposition of sentences which are to run consecutively.

It is extremely important that the security officer know the classificiations of crimes in the jurisdiction where he is employed. An officer could, for example, be guilty of false arrest if the crime committed were a misdemeanor, not committed in his presence, which he believed to be a felony. The actual sentence that is imposed for a crime in a court of law does not affect the classification of the crime. The classification is always based on the maximum sentence in those jurisdictions using this system of classification. As an example, the maximum penalty for robbery in a jurisdiction is five years confinement. Due to extenuating circumstances in the case, the offender is sentenced to only six months confinement. The crime that has been committed would still be classed as a felony.

If the security officer is to know that a felony has been committed, then he must know all the elements necessary to constitute the different felony classed crimes. Some of these crimes will be defined using either the common law definition or the definitions and elements which are used by the majority of jurisdictions. Once again, the security officer must be trained in the laws which govern his actions in the jurisdiction where he is employed.

CRIMINAL HOMICIDES

Homicide is the term given to the killing of one human being by another human being. Not all homicides are criminal. Some homicides are excusable and others are justifiable in the eyes of the law. Coverage of homicides will be restricted to those types which are classed as criminal.

Murder

To constitute murder, the homicide must have been the unlawful killing of a human being with malice. Murders generally fall into two categories or degrees. Any murder which has been willful, deliberate and premeditated or which has occurred during the attempt or actual commission of certain specified felony class crimes is murder in the first degree and is punished as the statutes provide. All other murders are classed as second degree murder and are also punished as the statutes provide.

Manslaughter

The crime of manslaughter is defined as the unlawful killing of a human being without malice. Manslaughter is usually divided into three distinct categories.

Voluntary Manslaughter. This is defined as the unlawful killing of a human being immediately following a sudden quarrel or which was done during the heat of passion. The difference between first degree murder and voluntary manslaughter often lies only in the time element between the provocation of the act and the act itself. For example, two men are involved in an argument which turns into violence where one man is being badly beaten by the other. The man on whom the beating is being inflicted pulls a gun and kills his antagonist. In that there does not appear that there was any time for conscious thought, the crime would be voluntary manslaughter. In the same situation, the man who is inflicting the beating feels that the other has had enough, stops beating him and starts to leave the area. In this case the man on whom the beating is being inflicted is angry about the beating he has been taking, pulls a gun, aims at his antagonist and kills him. The crime in this case could be murder but would probably be decided by the judge or jury hearing the case on the basis of whether the defendant had time to consciously plan to kill his antagonist. The premeditation does not require weeks, days or even hours but can amount to only seconds.

Involuntary Manslaughter.   This is defined as unlawful killing of a human being during the commission of an unlawful act, not amounting to a felony; or in the commission of a lawful act which might produce death, in an unlawful manner; or when done as the result of not exercising due caution or circumspection. The confusing part of this definition is how a lawful act can be committed that might produce death in an unlawful manner. What is being defined here is gross negligence. An example might be the operator of a ride at an amusement park. He is licensed and his ride has been inspected. He is aware that a weakness exists which was not detected and which could cause bodily harm or death should certain factors be present. He does not repair the weakness and several customers of the ride are killed. Providing that his knowledge of the deadly weakness could be shown, the operator could be charged with wrongful death under the involuntary manslaughter statute.

Motor Vehicle Manslaughter.   This is defined as the unlawful killing of a human being arising from the operation of a motor vehicle during the commission of an unlawful act, not amounting to a felony; the operation of a motor vehicle with gross negligence or during the commission of a lawful act, which might produce death in an unlawful manner, but without gross negligence. Some jurisdictions include motor vehicle manslaughter, and other types of accidental deaths which have been negligently caused, under a violation of the code identified as negligent homicide, which is generally considered to be a lesser included offense to the crime of manslaughter. The basic premise remains the same: the unlawful killing of one human being by another will be punishable as the statutes provide.

ASSAULT AND BATTERY

Assault may be defined as a threat of intended violence directed toward the person of another, which is coupled with the present ability to carry out the intent. A simple assault, without the presence of battery, would be classified as a misdemeanor.

Battery would be the action, following the assault, where a willful or intended touching took place, of one person by another or the willful touching of a person by some article which has been set into motion by the other. The striking of one person by another, where no grievous harm has been done, will normally be classed as a misdemeanor.

Felonious assault, assault with a deadly weapon, and aggravated assault are all terms used to describe assaults which are classed as felonies. The elements necessary for proving simple assault must be present but they must be coupled with the use of a deadly or dangerous weapon, the intent to inflict grievous bodily harm, the actual infliction of such harm, or the intent to commit murder.

ROBBERY

For a crime to be robbery it must be shown that there has been a taking of personal property, that the property was taken from the person or presence of the possessor, that the taking was against the will of the possessor, and that the taking had been accomplished through the use of force or fear. Any taking of property which does not contain all of the elements listed above would not constitute robbery. However, providing the taking was unlawful, the act would probably be classed as larceny.

BREAKING AND ENTERING OR BURGLARY

At common law the crime of burglary required that there be a breaking and entering, during the nighttime, of the dwelling place of another, for the purpose and with the intent of committing a larceny or felony within.

The crime of breaking and entering was used to cover crimes which did not meet all the elements of burglary. For example, a breaking and entering of any structure, at any time other than that covered under burglary, coupled with the intent to commit a crime within, normally constituted breaking and entering.

Today, most jurisdictions have statutes which define a burglary as the entering of a structure (usually defined in the penal code) with the intent to steal or to commit another felony. Burglary in itself is a felony and does not require that any other crime be committed following the entry as long as the intent to commit a crime within is proven. Other crimes which are committed during the course of a burglary would be separate and additional charges. As can be seen, the modern statues on burglary eliminate the need for the crime classification of breaking and entering.

Many jurisdictions do, however, have classes of burglaries. In those areas, first degree burglary would be the entry of a dwelling house which is occupied at the time that entry is made, entry during the nighttime, or entry by a person who is armed with a deadly weapon or who steals such a weapon during the course of the burglary. Second degree burglary would be most other types of burglaries. Some jurisdictions do have a third degree burglary, classified as a misdemeanor, for those cases where the charging of a felony is not appropriate. First and second degree burglaries are punishable as the statutes provide.

PROPERTY CRIMES

A security force will encounter property crimes much more frequently than perhaps any other category of crimes. They will be concerned with safeguarding the property of the employer, other employees working in the area and business visitors to the area. Therefore, the security supervisor must make sure that his security force personnel are particularly well versed in the elements required to prove that these crimes were committed.

Larceny

General.   To constitute the crime of larceny, it must be shown that there has been an actual or constructive taking, followed by the taking away, of the property of another. The taking must have been done without the consent and against the will of the owner, and there must be a willful and felonious intent, on the part of the thief, to permanently deprive the owner of his property.

From a Person.   All the elements of general larceny must be present and the taking must have been from the person or the immediate presence of the owner. This offense differs from robbery in that there is no requirement that the use of force or fear be an element of proof. Also, in a robbery the property can be taken from anyone who is in possession of it with no requirement that the person from whom it was taken is the owner of the property.

From a Building.   As above, all the elements of general larceny must be present and the taking would have to have taken place in a building, dwelling, office or other structure. In some jurisdictions, shop-lifting offenses would be prosecuted under this statute. Other jurisdictions, if not most, have specific statutes under which shoplifting offenses are prosecuted.

248

From a Motor Vehicle. All elements of general larceny must be present plus the taking must have been from the outside of a motor vehicle or from the inside of an unlocked motor vehicle. A larceny from the inside of a locked motor vehicle would be dealt with as burglary or breaking and entering, in most jurisdictions.

Malicious Destruction of Property

To have the elements which constitute this offense, it must be shown that the perpetrator has caused damage or injury to the personal property or structure (as defined in the statutes) of another and that the destruction was willful and malicious.

Property crimes, for the most part, are classified by the value of the property taken or damaged. In most jurisdictions the theft of property having a value of $100.00 or more would constitute a felony. Any theft or damage of property valued at less than $100.00 would be classed as a misdemeanor. Due to the escalating rate of inflation, many statutes are being revised to increase the value required to constitute a felony or grand larceny. These amounts are generally being raised to $250.00

THE LAWFUL SEARCH

A search of a person or place can be conducted lawfully in either of two ways, as outlined below:

1. The search of a person and his immediate surroundings subsequent to the making of a *lawful* arrest.
2. The search of a specified area pursuant to the issuance and service of a legally obtained search warrant.

In a search, following a lawful arrest, the immediate surroundings refer to that area under the immediate *control of* the arrestee. Any liberties which are taken, such as extending the search to areas other than where the arrest was made, will result in any evidence which is discovered being ruled inadmissible in a court of law. It also results in a shadow being placed over the credibility of the person making the search. However, if the suspect to be arrested is in flight and is observed dropping or throwing away an item suspected of being connected to the crime, the area of search may legally be extended to include the area where any item has been discarded during the flight.

To obtain a search warrant, an application must be made to a magistrate or judge. The application must be accompanied by an affidavit which details the reason for the search, the property which is believed to be in the location which is to be searched and the justification for the search. The justification must include the showing of reasonable cause for the belief that the property or evidence which is being sought is located in the place to be searched.

Any evidence of criminal activity that is found in the conduct of a lawful search, even that not connected with the crime or evidence which is the focal point of the search, will be admissible in a court of law to substantiate a charge which may be proven or substantiated by such evidence. Any evidence which is discovered, no matter how damaging, during the course of a search which has been ruled illegal is said to be fruit of the poisonous tree and will not be admitted as evidence in a court of law. The same rule applies to any evidence which is found or developed as a result of an illegally obtained admission or confession. The U.S. Supreme Court appears to be inclined to modify this rule, but as it presently stands such evidence is not admissible in court.

The security officer who has not been commissioned or deputized as a law enforcement officer is not obligated to advise a suspect or arrestee of his rights to remain silent or to have an attorney present during questioning. This duty is required of persons who are legally constituted officers of the law, but is not required of private.citizens. All security officers should be familiar with the legal rights of an accused, and suspects or arrestees should be advised of their rights by anyone making an arrest, even though it is not mandated by law. This, of course, would include security officers.

The rights referred to are guaranteed by the fifth and sixth Amendments to the Constitution of the United States, as defined by the Supreme Court in the cases of Miranda and Escobedo. They are basically as follows.

You have the right to remain silent. (Fifth Amendment guarantee against compulsory self incrimination). If you give up the right to remain silent, anything you say can and will be used against you in a court of law. You have the right to an attorney (Sixth Amendment guarantee), to have that attorney present during questioning; if you cannot afford an attorney, one will be appointed for you. Do you understand these rights as I have explained them to you? Do you wish to give up these rights at this time?

It should also be understood by security officers that the right of an individual to be protected against an unjust or unlawful search of his person or property is constitutionally guaranteed in the Fourth Amendment to the Constitution of the United States. Violations of constitutional rights can result in criminal charges and civil suits against the security officer and his employer.

This does not imply that the lunch box inspections or spot vehicle inspections which are conducted by security officers are illegal. These are considered inspections and are not searches in the legal sense. Evidence which is obtained during the course of such inspections is *not* admissible in court as evidence of a criminal act. It may, however, be grounds for the initiation of administrative action on the part of the company to terminate the services of the employee.

These inspections are designed to deter the removal of company property from the protected area and are not intended nor should they be used in any attempt to catch and prosecute a thief. If reasonable cause exists to show that an employee is removing stolen property from the protected area, an arrest should be made or a search warrant obtained prior to conducting a search for the suspected stolen property.

Security force personnel cannot be expected to be lawyers or to personally research the laws applicable to the performance of their duties. It is, therefore, the responsibility of the supervisor or risk manager to formulate policy pertaining to actions which they will authorize taken in specific situations. This policy must be based on the existing laws in the jurisdiction where they are employed. Under normal circumstances the making of arrests or the obtaining of search warrants, based on reasonable cause, will be left to local law enforcement personnel or company employed investigators. The security force *must* know and understand these laws in the event that outside assistance is not immediately available.

During initial and follow-up training sessions or on-the-job training, it should be emphasized that popular conceptions of what constitutes arrest and confinement, based on television shows and detective novels, are not usually the same as the interpretations given these actions by courts of law. It is surprising how many people, security officers included, believe that in order to make an arrest it is necessary to physically restrain a person or definitely state that he is being placed under arrest. Having these elements present would be useful in attempting to prove a charge of resisting arrest or escape. However, security officers should be made aware that all that is necessary to constitute an arrest is depriving a person of his freedom to come and go as he wishes.

Some courts have ruled, in false arrest proceedings, that a show of force or authority was sufficient to constitute an arrest even when the language used provided the detained person an option to leave if he so desired. An example of this might be two security officers flanking the suspect and stating "Would you like to come with us please?" The court's reasoning might be that the detained person was intimidated by the officers and did not relaize that he could exercise his freedom of movement and leave the area unmolested.

Imprisonment or confinement, much like arrest, is often misunderstood. These words seem to imply that a person is being locked into a cell. Actually, imprisonment under the law only requires denying a person's right to leave a place where he has been detained. A false imprisonment in some jurisdictions is treated as a degree of kidnapping. The place of imprisonment can be a sidewalk, an office, the counter of a store or anywhere else that a person may have been detained.

An arrest is unlawful, or false, any time the freedom of movement of an individual has been denied without due process or probable cause. Imprisonment is unlawful or false when an arrest has been effected which was not lawful or where due process was not followed.

The test of reasonable cause is stated as what the "reasonable man" would believe in like circumstances. If the "reasonable man" would believe that there was sufficient cause to make an arrest, then the arrest itself could be lawful even though subsequent events established that the arrested person had committed no crime.

Wrongful death was discussed in the chapter on firearms. It was stated that the only two incidents when a security officer was justified in using his weapon were in the protection of his own life or in the protection of the life of another person. A wrongful death action could be brought even if the death occurred while the officer was firing his weapon under either of the two circumstances stated as justified. An example might be a security officer at a banking facility following a robber out of the bank and being fired upon by the fleeing robber. His return fire would, under normal circumstances, be considered justified. However, if he were on a crowded street where his shooting would endanger other persons, his best course of action would be to stop the pursuit and not fire his weapon at the robber. If he did otherwise and a third party was killed as a result of his action, he could be the subject of a wrongful death action.

Wrongful deaths are not restricted to those brought about by the use of deadly weapons. A security officer on motorized patrol observes a vehicle speeding within the protected area and gives chase after illuminating his emergency lights and perhaps turning on a siren. During the high speed chase he loses control of his vehicle, smashes into a building and fatally injures several people. Even if a wrongful death action were not brought or sustained, the security force should be aware that their job is to protect personnel and property and not to place either in danger through their actions. In this situation it would be far better to notify the gates of the protected area of the description of the offending vehicle and to break off any attempt at pursuit which would be likely to endanger life or property. The goal is not only to acquaint security force personnel with their responsibilities under the law, but also, and primarily, to teach them to perform their duties using good judgment and common sense.

Security officers cannot be expected to take the proper actions unless they have been taught the actions that are available to them and which they can take with confidence. They should always be aware that they are contributing to the loss prevention process rather than causing it.

QUESTIONS FOR DISCUSSION

1. What is the difference between a misdemeanor and a felony in jurisdictions where the classification of crime is determined by the maximum penalty?
2. Under what circumstances may a security officer, with no special police authority, make a misdemeanor arrest?

3. What conditions must exist for a security officer to make an arrest of a suspected felon for a felony-type crime?
4. What elements are required to make a homicide a murder?
5. Discuss the main difference between the criminal acts of murder and manslaughter.
6. Under what circumstances may a lawful search of a person or property be conducted?
7. What is the main difference between the crimes of larceny and robbery?
8. Why is it necessary for a security officer to know the various elements that make up the different types of crimes?
9. Discuss the differences between the authority to make an arrest for a law enforcement officer as compared with a private citizen.
10. What are the conditions under which a security officer or other private citizen would be justified in making a felony arrest?

MAJOR CONSIDERATIONS

1. Security forces must be trained in the laws pertaining to arrest, confinement and search or seizure in the jurisdiction where they are employed if they are to be effective without incurring the risk of criminal or civil liabilities.
2. Policies and authorized actions must be spelled out for the security force, detailing what they may do and which actions require prior approval of supervisory personnel.
3. It is essential that security force personnel are well versed on the elements necessary to constitute criminal acts with which they may have to deal. They must know the difference between a misdemeanor and a felony and the actions they are authorized to take in dealing with them.

Chapter 27

Training Techniques

Instructing, in any course, requires the use of proven techniques to successfully teach the subject. However, before these techniques can be applied a thorough knowledge of the subject that is to be taught is necessary.

For the instructor the first step in teaching is getting the attention of his students. The first step, in this case, is probably the easiest. The second step is somewhat more difficult in that once the instructor has the attention of his students, if he is to be successful, he must keep it.

Picture yourself seated in a classroom awaiting the start of a class when a door bursts open and a man, holding a gun, begins shooting at the instructor. The instructor draws a gun from the podium and shoots the intruder who falls to the floor. The intruder gets up and flees from the room. Would you say that the instructor had succeeded in getting your attention? If your answer is yes, you could be only half right. The incident would certainly have the attention of the students in the room but it would not necessarily lead the attention of the class into the course of instruction that is to be given.

Opening a class with an attention getter, whether a joke, timely story or a staged event, is a good way to stimulate interest. The attention getter must relate to the subject matter of the class and should tend to direct the attention of the class, from their individual thoughts, to a collective focus on the subject.

An incident similar to the one described is adaptable and widely used in classes pertaining to the observation and description of persons involved in fast-moving incidents. Following the fleeing of the intruder from the classroom, the instructor would select members of the class to describe the intruder. He would be looking for identifying data such as height, weight, color of hair, skin coloring, type and color of clothing, type of weapon which was used and anything else that might aid in the later identification of the intruder. It is usually found that the untrained eye does not pick up this type of information; therefore it is impossible to obtain an adequate description of the intruder. This points up the need of the student for training in the techniques of observation and description and tends to focus the collective attention of the class on the material to be presented.

LESSON PLANNING

Classes should be planned to stimulate attention and to be kept sufficiently interesting to enable the students to grasp the subject matter being presented. The more interest that can be generated, the greater the rate of retention of the lesson.

If a subject is to be taught over more than four periods of instruction, an outline should be prepared to designate what elements of the subject will be taught in each period of instruction.

To assist in instructing a class, a lesson plan should be developed to be used as a guide while instructing. A good lesson plan should contain the following information in a format similar to the one shown. It is recommended that a lesson plan cover no more than four hours of instruction. A separate lesson plan for each hour to be presented would be preferred.

SUBJECT

Hour_____of_____Hours

I.   Introduction: 10 minutes
    A. Objectives: This section should detail what is to be accomplished by learning the material to be presented, why it is important to the student, where it fits in his daily routine or job performance and should contain the attention-getting anecdote, timely story or staged event.
    B. Subject: List the material, pertaining to the subject, which will be presented during this block of instruction.

II. Body: 30 minutes—Outline major points to be discussed, for example, if the subject is perimeter security.
    A. Type of barrier.
       1. Chain link fence.
       2. Block wall.
       3. Building walls.
       4. Other.
    B. Type of gates
       1. Pedestrian
          a. Turnstile.
          b. Chain link.
          c. Other.

2. Vehicular.
   a. Overhead.
   b. Swinging cantilever.
   c. Other.
C. Providing area clearance
   1. Trees.
   2. Trash and debris.
III.   Review: 5 minutes—Conduct a brief review of the material which was covered in this block of instruction.
IV.   Summary: 5 minutes—Summarize the major points of importance to be remembered.
V.   Critique: 5 minutes—Determine if there are any questions in the minds of the students on any of the material which has been covered.
Break: 10 minutes

Lesson plans are teaching *guides*. As such, they should not be detailed but should contain only the major points to be covered. They are to be used to insure that all pertinent material is covered in an orderly manner.

TEACHING AIDS

To properly emphasize major points of subject material while maintaining class interest, timely stories, jokes or staged events can be interwoven throughout the period of instruction. The best method of accomplishing this is by relating an interesting occurrence, involving the subject of the class, from the teacher's own experience or knowledge. The use of jokes or staged events which give emphasis to the subject of discussion are also effective. It is not intended to imply that a major part of the period of instruction should be taken up with the telling of "war stories", jokes or entertaining the class through many staged events. These are only methods to stimulate and retain interest and must be included in the class in a proper balance.

A good instructor, when planning a class, will review slides, training films, text or reference books, mockups or any other available training aids for possible inclusion in the class. Training aids should be selected which will best provide the information that is to be presented. Such aids to the training program should be integrated into the class time as advantageously as possible.

It is not recommended, for example, that three hours of lecture be followed by three hours of films, followed by three more hours of lecture. It would be much better to blend the movies into the lecture periods in such a way as to keep class interest at the highest level.

For an instructor to be really effective he should always attempt to appeal to as many of the five senses as possible. That is, the class should receive the information by hearing it explained, while seeing it displayed through the use of blackboards, diagrams, photographs or demonstrations. The sense of touch can be added by providing each student with the object, if any, being discussed. For example, in a weapons care and maintenance class, each student would be given a weapon to use as he follows the instruction.

Probably one of the easiest subjects to teach would be how to bake a chocolate cake. The class could hear the recipe and see it mixed. Thereafter they would be able to smell it baking, hold it in their hand and taste the results.

STUDENT PARTICIPATION AND ATTENTION

Student participation should be encouraged by asking questions, calling upon students for opinions, personal experiences or knowledge and by urging the students to ask questions. Questions should be used to insure that students have understood the subject matter that has been presented. Questions are also valuable in having the material repeated for all to hear once again. The more often the material can be repeated, the greater the possibility that it will be retained.

In maintaining the attention of the class, the instructor will have to make every effort to minimize student discomfort and distractions. The instructor should not be someone whose other duties will require frequent interruptions of the class while he answers questions, makes decisions or is called upon for assistance. From personal experience, it is extremely difficult to keep the attention of a class being held in an outdoor setting during a pelting rainstorm. The use of plush upholstered couches or chairs is not advocated since too much comfort can be as much a problem as too little. Wherever possible, however, the class setting should be sufficiently comfortable so as not to be distracting.

EVALUATING STUDENTS AND THE TRAINING PROGRAM

Students should be required and allowed to practice what they have been taught by being given the opportunity to put their newly acquired knowledge to work. One example is a sketch or diagram of a site which can be an industrial, institutional or other type facility, showing the buildings and grounds, identifying building entrances, parking areas, shipping/receiving docks (if applicable), office spaces, restricted areas, production or storage areas and so on.

Using this inexpensive training aid, which can be prepared locally, allows the students to plan and complete the security. The following example uses an industrial facility. As the students answer the following questions, their solutions can be discussed and the resulting solution may be plotted on the drawing or diagram.

1. Where would you install perimeter barriers? What type barrier would be used? What type gates would be used and where would they be placed?
2. Where would guard posts be established?
3. What type of identification system and movement control procedures would be established and in which areas would more stringent requirements be needed?
4. Which entrances would provide the best control for the location of employee entrances?
5. Would you install closed circuit television? If so, where would you place the camera positions? Why?
6. If closed circuit television is recommended, which positions would require pan, tilt and zoom lens capabilities?
7. Would a video tape capability be recommended? On all camera positions or only selected positions? Discuss.
8. Which areas of the facility would require special consideration in the installation of a protective lighting system?
9. How would you identify and control vehicles entering the parking areas? The protected area?
10. Would you anticipate any special problems with traffic entering the facility or exiting onto main traffic arteries during shift changes?
11. Would you install alarms? If so, what type and where?
12. What method would you select to monitor an installed alarm system? Central station or proprietary? Why?
13. Where would you establish patrol routes? In which areas, if any, would you require patrol supervision stations?
14. What type of supervision would you select to insure the integrity of the patrols?
15. Would you recommend foot or motorized patrols? A combination of these types? What type of equipment would be required to make the patrols effective?
16. Is there a need for special fire protection equipment in any of the areas discussed? What would you recommend?
17. Are there any other considerations you would need to know to properly secure this facility?
18. Having secured the area, what would be the manpower requirements of the security force presuming a 24-hour day, 6-day per week work schedule for plant operations? Have you allowed for the difference in manpower requirements during operational and non-

operational hours? Have you properly staffed all operating gates and patrol functions? Have you provided for alarm and closed circuit television response? Have you considered a reserve or backup force? Have you provided for supervisory personnel?

Using this or a similar method, effectiveness of the training program can be reviewed by determining how much the students have learned and retained. At the same time, one more review of the important lessons will be provided to reinforce the learning and retention process.

After all classroom and practical instruction has been completed, if it is economically possible, a period of on-the-job training should be devised. This training is best accomplished by having each student work each position being manned within the protected area, under the guidance of an experienced officer. Daily critiques would be useful to insure that the students are receiving and absorbing the required training.

Following formal training sessions, students are asked to complete a critique of the presentation, class format and the information provided. The critique may or may not be signed, as the student prefers, but needs to contain his opinion of the quality and quantity of the instruction received. Criticism and suggestions for improvement are solicited. By obtaining this type input, materials and methods can be continuously evaluated and improved.

Examinations, if desired, may be oral, written, practical exercises or a combination of all three. Examinations and critiques are given primarily as one more reinforcement of the teaching process. They point out whether the student has received any misconceptions of his duties or his responsibilities and is able to correct them. In this manner the quality of the training as well as the training program can be improved.

Providing an excellent initial training program for the security force is essential in establishing a professional force. Training does not and cannot stop with the initial training. Ongoing training programs must be established and implemented to provide new or changing procedures and to insure that the security force does not become complacent, develop bad habits or forget the lessons leaned in the initial program.

SAMPLE COURSE FORMAT

It is impossible to outline a training program suitable for all the different security programs and situations that will be encountered. Several general formats for initial and follow-up training programs will be provided, which will have to be adapted to the individual needs or circumstances of the different risk managers and supervisors. It is understood that time and financial considerations will play major roles in determining the extent and content of the program which is ultimately adopted. Except in the largest security departments or when setting up a new department, the number of personnel being trained at any given time will probably be minimal and would also have an effect on the type and extent of formal training provided. The hours which are recommended for teaching a particular subject are flexible and should be adjusted depending upon the needs and duties of the individual department.

In many cases it may be desirable to give only introductory training in the general security subjects while alloting more time to the specialized knowledge needed for the performance of the officer's duties. Time allotted for perimeter security, area security, shipping/receiving operations or weapons training would be minimal or could be eliminated for security forces assigned to financial institutions, retail stores, high rise buildings, or unarmed posts. Each supervisor will have to evaluate the subjects that need to be included in the training program for security officers under his supervision.

The following format is based on an 80-hour course of instruction. It can be used as a total two-week initial training program or the security supervisor can select those subjects most important to his program for inclusion in a formal training program and schedule other periods into smaller blocks for a continuing weekly training program. In departments where weapons are not used or planned, the time allotted to weapons training can be utilized to provide the extremely important specialized training required for that particular department. The hours ascribed to each subject should not be considered iron cast but should be adjusted by the needs and duties of each department.

WEEK 1

MONDAY

8:00 A.M. — 12:00 P.M.

INTRODUCTION (4 Hours)
1. History and purpose of security force. 2 Hours
2. Overview of facility operations. 2 Hours

12:00 P.M. — 1:00 P.M.

LUNCH

1:00 P.M. — 5:00 P.M.

SECURITY HAZARDS (4 Hours)
1. Human Hazards. 2 Hours
2. Natural Hazards. 2 Hours

TUESDAY

8:00 A.M. — 12:00 P.M.

LOSS PREVENTION (4 Hours)
1. General. 1 Hour
2. Specific problems and methods unique to each facility. 3 Hours

12:00 P.M. — 1:00 P.M.

LUNCH

1:00 P.M. — 5:00 P.M.

IDENTIFICATION AND CONTROL (4 Hours)
1. Personnel. 2 Hours
2. Vehicles. 2 Hours

WEDNESDAY

8:00 A.M. — 12:00 P.M.

KEYS AND LOCKING DEVICES (4 Hours)
1. Key control. 1 Hour
2. Locks, uses, types, vulnerability. 3 Hours

12:00 P.M. — 1:00 P.M.

LUNCH

1:00 P.M. — 5:00 P.M.

FIRE PREVENTION (4 Hours)
1. Elements and classes of fires. 2 Hours
2. Types of fuel and prevention actions. 2 Hours

THURSDAY

8:00 A.M. — 12:00 P.M.

FIRE EQUIPMENT (4 Hours)
1. Standpipe hoses. 1 Hour
2. Sprinkler systems. 1 Hour
3. Extinguishers. 2 Hours

12:00 P.M. — 1:00 P.M.

LUNCH

1:00 P.M. — 5:00 P.M.

FIRE EQUIPMENT (4 Hours)
1. Practical exercises in use of portable extinguishers and hoses.
    4 Hours.

FRIDAY
8:00 A.M. — 12:00 P.M.

FIRST AID (4 Hours)
1. As prescribed by ARC. (4 Hours)

12:00 P.M. — 1:00 P.M.

LUNCH

1:00 P.M. — 5:00 P.M.

FIRST AID (cont'd) (4 Hours)

WEEK 2

MONDAY

8:00 A.M. — 12:00 P.M.

SAFETY (2 Hours)

DISASTER AND EMERGENCIES (2 Hours)*

12:00 P.M. — 1:00 P.M.

LUNCH

1:00 P.M. — 5:00 P.M.

LEGAL AUTHORITY (4 Hours)

*Will require augmentation during follow-up and on the job training.

TUESDAY

8:00 A.M. — 12:00 P.M.

ACCESS SYSTEMS (1 Hour)

PUBLIC RELATIONS (1 Hour)

REPORT WRITING (2 Hours)*

12:00 P.M. — 1:00 P.M.

LUNCH

1:00 P.M. — 5:00 P.M.

PHYSICAL SECURITY (4 Hours)

1. Perimeter security. 2 Hours
2. Area security. 1 Hour
3. Building security. 1 Hour

*Will require continuous review and monitoring.

WEDNESDAY

8:00 A.M. — 12:00 P.M.

FIREARMS TRAINING (4 Hours)
1. Safety. 2 Hours
2. Care and cleaning. 2 Hours

12:00 P.M. — 1:00 P.M.

LUNCH

1:00 P.M. — 5:00 P.M.

FIREARMS TRAINING (cont'd) (4 Hours)
1. Position and grip. 2 Hours
2. Sight alignment and trigger squeeze. 2 Hours

THURSDAY

8:00 A.M. — 5:00 P.M.

FIREARMS TRAINING (cont'd) (8 Hours)
1. Dry fire exercises. 2 hours
2. Practice firing. 2 hours
3. Qualification firing. 4 hours

NOTE: In lieu of firearms training for security forces which are not armed, familiarization tours of the protected facility or area are recommended, with emphasis on safety, emergency evacuation routes and emergency procedures.

FRIDAY

8:00 A.M. — 12:00 P.M.

PROTECTIVE LIGHTING (2 Hours)

ALARM SYSTEMS (2 Hours)

12:00 P.M. — 1:00 P.M.

LUNCH

1:00 P.M. — 5:00 P.M.

CLOSED CIRCUIT T.V. (1 Hour)

EXECUTIVE PROTECTION (1 Hour)

EXAMINATION AND CRITIQUE (2 Hours)

In addition to the above program, shift supervisors or one designated person on each shift should be given the advanced 40-hour course on first aid. Patrol techniques should be covered in classes on safety, physical security, protective lighting, fire prevention and emergency procedures. Bomb threats and civil disturbances, if required by the security force being trained, can be given in classes pertaining to emergency procedures.

If the industry is to gain true professionalism, minimum standards of training and education will have to be established and enforced. The establishment of the Certified Protection Professional (CPP) for security/risk managers has been a giant step in the right direction. It cannot, however, stop there. The supervisor and other security force personnel must be included by setting up goals and standards to be achieved at various levels of competence within the industry.

QUESTIONS FOR DISCUSSION

1. What is the first thing an instructor must master before attempting to teach?
2. What is the purpose of using an attention getter?
3. What information should be contained in a lesson plan?
4. When should a lesson outline be prepared and what is its purpose?
5. What are some methods of maintaining class interest while not deviating from the subject?
6. Discuss the elements necessary for the preparation of a class.
7. In what manner can the information you want to impart be best presented to insure that it is understood and retained?
8. How effective is student participation in a class?
9. Must tests or examinations be reduced to writing and be graded in order for them to be effective?
10. Once an effective initial training program has been conducted, is it necessary to conduct periodic follow-up training? Why?

MAJOR CONSIDERATIONS

1. For instruction to be effective, the instructor must know his subject, plan each period effectively and keep the interest and attention of the students.
2. The objective of teaching is that the student understand, retain and be able to use the subject matter presented.
3. An effective training program does not stop with the initial formal training; it must be continuously reviewed and updated in a never-ending cycle.

Epilogue

No single work will ever fully cover the multitude of duties which are encountered by the members of the security profession. It is our hope that we have been able to provide a basic understanding of the changing concepts of security.

We would like to see the product of our efforts used to expand the knowledge and importance of the security professional. We have known and worked with security officers throughout this country as well as in Europe, Central and South America and the Far East. We have learned a great deal from these associations and sincerely hope we have succeeded in passing on some of this knowledge in the pages of this book.

To those of our associates from whom we have already learned, we give our warmest thanks. To those persons from whom we expect to learn in the future, we thank you also.

Selected Services
and Products

The following companies and organizations are listed for the convenience of the reader. They represent some of the products and services mentioned in this book. We do not imply any recommendation or endorsement of the products or services offered. However, all are considered reputable businesses. Conversely, the failure to list other firms engaged in providing similar goods or services does not imply that they are not reputable. There is not sufficient space on these pages to list all firms engaged in supplying security products and services.

As stated in several of the chapters of this book, it is recommended that the potential consumer compare the products and services available in his area, the supervision or maintenance response that is available and select the one that most nearly meets his needs.

- Access Control Systems, 2105 South Hardy Drive, Tempe, AZ 85282. (Design, installation and maintenance of access control systems).
- ADT, One World Trade Center, New York, NY 10048 (Central Station Alarms).
- Armored Vehicle Builders, P.O. Box 62, Pittsfield, MA 01201 (Armored Cars).
- Best Lock Company, P.O. Box 50444, Indianapolis, IN 46250 (Locking Devices).
- E.J. Brooks, 164 North 13th Street, Newark, NJ 07107 (Seals and Locking Devices).
- Burke Security, 880 Third Avenue, New York, NY 10022 (Security Forces).
- Burns International Security Services, Inc., 320 Old Briarcliff Road, Briarcliff Manor, NY 10510 (Consulting, Electronics, Guard Forces, Investigations).

- California Plant Protection, 6727 Odessa Avenue, Van Nuys, CA 91406 (Guard Forces).
- Cardkey Systems, 20339 Nordhoff Street, Chatsworth, CA 91311 (Access Control Systems).
- Cushman, P.O. Box 82409, Lincoln, NE 68501 (Police and Utility Vehicles).
- Detex, 4147 Ravenswood Avenue, Chicago, IL 60613 (Watchclocks and Alarm Systems).
- Diebold, Inc. 1550 Grand Boulevard, Hamilton, OH 45011 (Security Cabinets and Vaults).
- General Electric Company, CCTV Operations, Owensboro, KY 42301 (Closed Circuit Television)
- Globe Security Systems, 2503 Lombard Street, Philadelphia, PA 19416 (Security Forces).
- Guardsmark, Inc., 22 South Second Street, Memphis, TN 38103 (Consulting, Guard Forces, Investigations, Training).
- Honeywell Fire and Safety Systems, Dept. 06723, Honeywell Plaza, Minneapolis, MN 53408 (Alarm and Access Control Systems).
- Infinetics, Inc., P.O. Box 2330 Wilmington, DE 19899 (Metal Detectors).
- Identatronics, 425 Lively Boulevard, Elk Grove Village, IL 60007 (Identification Cards and Processing Systems).
- Kidde Belleville, 675 Main Street, Belleview, NJ 07109 (CCTV, Fire and Intrusion Alarm Systems).
- Mardex Videoguard, 900 Stierlin Road, Mountain View, Ca 94043 (Closed Circuit Television and Access Control Systems).
- Master Lock Company, 2600 North 32nd Street, Milwaukee, WI 53210 (Locking Devices — Padlocks).
- Mosler, Dept. SC-80, 1401 Wilson Boulevard, Arlington, VA 22209 (Security Cabinets and Vaults).
- MTI Teleprograms Inc., 4825 North Scott Street, Suite 23, Schiller Park, IL 60176 (Audio Visual Training Programs).
- National Fire Protection Association, 470 Atlantic Avenue, Boston, MA 02210 (Fire Protection Standards).
- National Rifle Association 1600 Rhode Island Avenue, Washington, D.C. 20036 (Weapons Training and Allied Products).
- NEC America Inc., 130 Martin Lane, Elk Grove Village, IL 60007 (Closed Circuit Television Systems).
- Norton Company, Air Space Devices, P.O. Box 7500, Cerritos, CA 90701 (Perimeter Alarm Systems).
- Panasonic Company, Video Systems Division, Dept. SMT-780, One Panasonic Way, Secaucus, NJ 07094 (Closed Circuit Television Systems).

- Par-Kut International, Inc. 25500 Joy Boulevard, Mount Clemens, MI 48045 (Gate and Guard Houses).
- Phillips Electronic Instruments, Inc., Security Systems Group, 85 McKee Drive, Mahwah, NJ 07430 (X-Ray Units, Explosives and Metal Detectors).
- Pinkertons, Inc., 100 Church Street, New York, NY 10007 (Consulting, Investigations, Security Forces).
- RCA, CCVE Marketing, New Holland Avenue, Lancaster, PA 17604 (Closed Circuit Television Systems).
- Rusco Electronic Systems, 1840 Victory Boulevard, Glendale, CA 91201 (Access Control Systems).
- Sargent & Company, 100 Sargent Drive, New Haven, CT 06509 (Locking Devices).
- Schlage Electronics, 1135 East Argues Avenue, Sunnyvale, CA 94086 (Locking Devices and Access Control Systems).
- Sensor Engineering Company, 2155 State Street, Hamden, Ct 06517 (Access Control Systems).
- Shorrock, Inc., Parkway Industrial Center, 7235 Standard Drive, Hanover, MD 21076 (Access Control and Alarm Systems).
- Simplex Security Systems, Inc., 78 Front Street, Collinsville, CT 06022 (Locking Devices).
- The Stanley Works, 5740 East Nevada, Detroit, MI 48234 (Electronic Gate Systems).
- Vicon Industries, Inc., 125 East Bethage Road, Plainview, NY 11803 (Closed Circuit Television Accessories).
- Wackenhut Corporation, 3280 Ponce de Leon Boulevard, Coral Gables, FL 33134 (Security Forces).
- Wells Fargo Guard Services, 1633 Littleton Road, Parsippany, NJ 07054 (Security Forces).
- Wilson Electronics, P.O. Box 19000, Las Vegas, NV 89119 (Radio Communications Equipment).

Selected Bibliography

American Red Cross, *Advanced First Aid & Emergency Care*, Doubleday & Company, Garden City, N.Y., 1973.

American National Standard Practice For Protective Lighting, RP-10, Illuminating Engineering Society, New York, N.Y., 1970.

Anderson, Edward J., "A Study of Industrial Espionage," Parts I & II, *Security Management*, January & March 1977.

Astor, Saul D., "Contract Guards: The Facts As I See Them," *Security World*, July/August 1975.

Astor, Saul D., *Loss Prevention Controls and Concepts*, Butterworth Publishers, Woburn, Mass., 1978.

Bassiouni, M. Cherif, *Citizens Arrest*, Charles C. Thomas, Springfield, Ill., 1977.

Burstein, Harry, *Hotel Security Management*, Praeger Publishers, New York, N.Y., 1975.

Burstein, Harry, *Industrial Security Management*, Praeger Publishers, New York, N.Y., 1977.

Cohen, Joseph, "Choosing Contract or Proprietary Security," *Security Management*, October 1979.

Bruyne, P. de, "Developments in Signature Verification," *Security Management*, January 1979.

Finneran, Eugene D., "Loss Potential, Risk Should Decide Security Budget," *Computerworld*, August 27, 1975.

Fisher, James A., *Security for Business and Industry*, Prentice-Hall, Englewood Cliffs, N.J., 1979.

Green, Gion and Raymond C. Farber, Introduction to Security, Butterworth Publishers, Woburn, Mass., 1969.

Healy, Richard J., *Design for Security*, John Wiley & Sons, New York, N.Y., 1968.

Higgens, Clay E., "The Security Director As Trainer," *Security Management*, May 1980.

Jackson, Barbara Bund, "Manage Risk In Industrial Pricing," *Harvard Business Review*, July-August 1980.

Kakalik, James S. and Sorrell Wildhorn, *The Private Police, Security and Danger,* Crane Russak & Company, New York, N.Y., 1977.

Kakonis, Tom E. and Donald H. Hanzek, *Police Report Writing,* Gregg Division, McGraw Hill Book Company, New York, N.Y., 1976.

Kelly, Robert E., "Should You Have An Internal Consultant?" *Harvard Business Review*, November-December 1979.

Knowles, Graham, *Bomb Security Guide,* Butterworth Publishers, Woburn, Mass., 1976.

Kupperman, Robert and Darrell Trent, *Terrorism—Threat, Reality, Response,* Hoover Institution Press, Stanford, Calif., 1979.

Kuhns, Roger, "Photographic Identification," *Security Management,* March 1978.

Lapides, G.A., "Exit Interviews As A Loss Prevention Technique," *Security Management,* May 1979.

Lewis, Robert A., "Better Security Through LPS Lighting," *Security Management,* April 1978.

Lipman, Ira A., *How To Protect Yourself From Crime,* Atheneum, New York, N.Y., 1975.

Lipman, Mark, *Stealing,* Harpers Magazine Press New York, N.Y., 1973.

Melnicoe, William and Jan C. Mennig, *Elements of Police Supervision,* Glencoe Press, Beverly Hills, Calif., 1970.

National Advisory Committee on Criminal Justice Standards and Goals, Private Security, *Report of the Task Force on Private Security,* Government Printing Office, Washington, D.C., 1976.

National Fire Protection Association Publications, NAFP, Boston, Mass., Continuous updating.

National Rifle Association Publications and Training Aids, Washington, D.C. Continuous updating.

Newman, Oscar, *Defensible Space,* MacMillan Publishing Company, New York, N.Y., 1973.

Perkins, Rollin, M. *Criminal Law and Procedure,* 3rd Ed., The Foundation Press, Brooklyn, N.Y., 1966.

U.S. Congress, *Occupational Safety and Health Act,* Government Printing Office, Washington, D.C., 1970.

Oliver, E. and J. Wilson, *Practical Security in Commerce and Industry,* London, 1979.

Post, Richard S. and Arthur A. Kingsbury, *Security Administration: An Introduction*, Charles C. Thomas, Springfield, Ill., 1977.

Prosser, William A., *Law of Torts,* West Publishing Company, St. Paul, Minn., 1955.

Prosser, William A. and Young B. Smith, *Torts,* 4th Ed. The Foundation Press, Brooklyn, N.Y., 1967.

Pulcini, John V., "Cost Effectiveness in Access Control," *Security Management,* November 1977.

Reber, Jan R. and Paul Shaw, *Executive Protection Manual,* Motorola Teleprograms, Schiller Park, Ill., 1976.

Reeser, Clayton and Marvin Loper, *Management, The Key To Organizational Effectiveness,* Scott Foresman and Company, Glenview, Ill., 1978.

Sasser, W. Earl Jr. and Frank S. Leonard, "Let First-Level Supervisors Do Their Job," *Harvard Business Review,* March-April 1980.

Scott, William H., "Vehicle Ambush In The U.S." *Security Industry and Product News,* August 1979.

Sennewald, Charles, *Effective Security Management,* Butterworth Publishers, Woburn, Mass., 1978.

Singer, Lloyd and Jan Reber "A Crises Management System," *Security Management,* September 1977.

Strobl, Walter M., *Crime Prevention Through Physical Security,* Marcel Dekker New York, N.Y., 1978.

Strobl, Walter M., *Security,* Strobl Security Services, Houston, Tex., 1973.

Superintendent of Documents, *Crimes Against Business: A Management Perspective,* Government Printing Office, Washington, D.C.

U.S. Department of the Army, *Civil Disturbances,* Government Printing Office, Washington, D.C.

U.S. Department of the Army, *Physical Security,* FM 19-30, Government Printing Office, Washington, D.C., 1971.

Walsh, T.J. and R.J. Healy, *Protections of Assets Manual,* The Merritt Company, Santa Monica, Calif., Continuous updating.

Weber, Howard T., "Implementing A Practical Risk Management Program," *Security Managment,* July 1980.

Wels, Byron, *Fire and Theft Security Systems,* Tab Books, Blue Ridge Summit, Pa., 1971.

Weston, Paul B. and Kenneth M. Wells, *Criminal Investigation—Basic Procedures,* Prentice-Hall, Englewood Cliffs, N.J., 1979.

Whitehurst, Susan A., "Update On Fire Protection," *SDM,* March 1980.

Williams, H.E. "Undercover Investigations," *Security Management,* September 1977.

Williams, Mason, *The Law Enforcement Book of Weapons, Ammunition and Training Procedures,* Charles C. Thomas, Springfield, Ill., 1977.

INDEX